CAMBRIDGE LIBRARY COLLECTION

Books of enduring scholarly value

Technology

The focus of this series is engineering, broadly construed. It covers technological innovation from a range of periods and cultures, but centres on the technological achievements of the industrial era in the West, particularly in the nineteenth century, as understood by their contemporaries. Infrastructure is one major focus, covering the building of railways and canals, bridges and tunnels, land drainage, the laying of submarine cables, and the construction of docks and lighthouses. Other key topics include developments in industrial and manufacturing fields such as mining technology, the production of iron and steel, the use of steam power, and chemical processes such as photography and textile dyes.

Lighthouse Construction and Illumination

Thomas Stevenson (1818–87) was the son of the engineer Robert Stevenson, and father of the writer Robert Louis Stevenson. Like his brothers David and Alan, he became a lighthouse designer, being responsible for over thirty examples around Scotland. Throughout his career he was interested in the theory as well as the practice of his profession, and published over sixty articles on engineering and meteorology. He was an international expert on lighthouses, and advised on projects in India, China, Japan, New Zealand and Canada. *Lighthouse Construction and Illumination*, published in 1881, was an expanded version of his *Lighthouse Illumination* of 1859, and remained a standard text. Many of his improvements in illumination came into universal use. According to his son, as a result of Thomas's work 'the great sea lights in every quarter of the world now shine more brightly'.

Cambridge University Press has long been a pioneer in the reissuing of out-of-print titles from its own backlist, producing digital reprints of books that are still sought after by scholars and students but could not be reprinted economically using traditional technology. The Cambridge Library Collection extends this activity to a wider range of books which are still of importance to researchers and professionals, either for the source material they contain, or as landmarks in the history of their academic discipline.

Drawing from the world-renowned collections in the Cambridge University Library, and guided by the advice of experts in each subject area, Cambridge University Press is using state-of-the-art scanning machines in its own Printing House to capture the content of each book selected for inclusion. The files are processed to give a consistently clear, crisp image, and the books finished to the high quality standard for which the Press is recognised around the world. The latest print-on-demand technology ensures that the books will remain available indefinitely, and that orders for single or multiple copies can quickly be supplied.

The Cambridge Library Collection will bring back to life books of enduring scholarly value (including out-of-copyright works originally issued by other publishers) across a wide range of disciplines in the humanities and social sciences and in science and technology.

Lighthouse Construction and Illumination

Thomas Stevenson

CAMBRIDGE UNIVERSITY PRESS

Cambridge, New York, Melbourne, Madrid, Cape Town, Singapore,
São Paolo, Delhi, Dubai, Tokyo, Mexico City

Published in the United States of America by Cambridge University Press, New York

www.cambridge.org
Information on this title: www.cambridge.org/9781108026895

This edition first published 1881
This digitally printed version 2010

ISBN 978-1-108-02689-5 Paperback

LIGHTHOUSE CONSTRUCTION

AND

ILLUMINATION

LIGHTHOUSE CONSTRUCTION

AND

ILLUMINATION

BY

THOMAS STEVENSON, F.R.S.E., F.G.S.

MEMBER OF THE INSTITUTION OF CIVIL ENGINEERS;

AUTHOR OF 'LIGHTHOUSE ILLUMINATION,' 'THE DESIGN AND CONSTRUCTION OF HARBOURS,' ETC.

LONDON AND NEW YORK

E. & F. N. SPON

MDCCCLXXXI

Printed by R. & R. Clark, *Edinburgh*.

TO

Sir John Hawkshaw,

F.R.SS. L. & E., F.G.S., ETC.

PAST PRESIDENT OF THE INSTITUTION OF CIVIL ENGINEERS.

MY DEAR HAWKSHAW,

I beg leave to dedicate to you the following pages, not only on account of the distinguished position which you justly hold in the profession, but as a mark of my great personal regard and friendship.

Very truly yours,

THOMAS STEVENSON.

EDINBURGH, *December* 1880.

PREFACE.

In the following pages I have done my best to give a condensed statement, 1*st,* Of the facts and principles which regulate the design and construction of Lighthouse Towers in exposed situations; and 2*d,* Of the practical application of optics to Coast Illumination. The history of the introduction of the optical agents, and their combinations to meet different requirements, have, so far as consistent with a clear exposition and division of the subject, been arranged in the chronological order of their publication, or of their employment in Lighthouses.

Though there are many excellent publications on special branches of this very important application of optical science, yet a systematic treatment of the whole subject, brought up to the present date, seems to be necessary. This will appear from the fact that so many new optical agents and modes of dealing with the light have been invented since the introduction of the admirable dioptric system of Fresnel, which consisted of only four agents, and three forms of apparatus in which these agents were combined.

All the later improvements have been fully described and illustrated, so as to enable the engineer to select from the different designs, such plans as will best suit the special peculiarities of any line of coast which he may be called on to illuminate.

EDINBURGH, *December* 1880.

CONTENTS.

CHAPTER I.

CHAPTER II.

CHAPTER III.

CHAPTER IV.

CHAPTER V.

CHAPTER VI.

CHAPTER VII.

CHAPTER VIII.

CHAPTER IX.

CHAPTER X.

APPENDIX.

CHAPTER I.

LIGHTHOUSE CONSTRUCTION.

1. Among the many works of man which prove the truth of the saying that "knowledge is power," we must not omit those solitary towers, often half-buried in the surge, that convert hidden dangers into sources of safety, so that the sailor now steers for those very rocks which he formerly dreaded and took so much care to avoid. *Olim periclum nunc salus* would be a fitting inscription for all such beacons of the night. It was therefore a true, as well as a noble saying of Louis XIV., when one of his privateers seized the workmen at the Eddystone, and took them to France in hope of reward, that "though he was at war with England, he was not at war with mankind." It is to the history and peculiarities of construction, and the general principles which have to be kept in view in designing such towers, that we have now to devote our attention.

In considering this subject, the first and most important matter requiring our attention, is the element with which we have to deal, its destructive power when excited by storms, and the directions of the forces it exerts against the masonry of a tower placed within its reach.

In his History of the Eddystone Lighthouse Smeaton says—"When we have to do with, and to endeavour to control, those powers of nature that are subject to no calculation, I trust it will be deemed prudent not to omit, in such a case, anything that can without difficulty be applied, and that would be likely to add to the security." This statement of our greatest maritime engineer indicates the propriety of carefully collecting any facts that may help us to a more accurate estimation of those forces which he regarded as being "*subject to no calculation.*" Now, it must be recollected that it is not an agent of constant power that we have to deal with here, for the towers which are built on rocks in the sea are not all assailed equally. The forces that act against buildings confronting the open ocean are never found in seas of smaller area, where, however, they may yet prove sufficiently powerful to destroy our works, unless we have correctly estimated the amount of exposure. In order to have a safe guide to direct us in each case, we must know, approximately at least, the heights of the waves which are due to different lengths of "fetch." Before making any design, therefore, we ought, if possible, to know the law of generation of waves in relation to the distance from the windward shore.

2. *Law of the Ratio of the Square Roots of the Distances from the Windward Shore.*—In the Edinburgh Philosophical Journal for 1852 I stated, as the result of a series of observations begun in 1850, that the height of waves increased most nearly "in the ratio of the square roots of their distances from the windward shore." Where h is the height of waves in feet, and d the fetch in miles, the formula $h = 1{\cdot}5\sqrt{d}$, from which the storm curve was laid down, will be found sufficiently accurate for all but very short fetches. Where the fetch is less than 39 miles, the most accurate formula is $h = 1{\cdot}5\sqrt{d} + (2{\cdot}5 - \sqrt[4]{d})$. In calculating the annexed table, both formulæ were employed.

The ordinates representing the height of waves in heavy

gales (Fig. 1) were all that had been observed when the diagram was made, but since then many more have been obtained, which also corroborate the law. The abscissæ show the lengths of fetch.

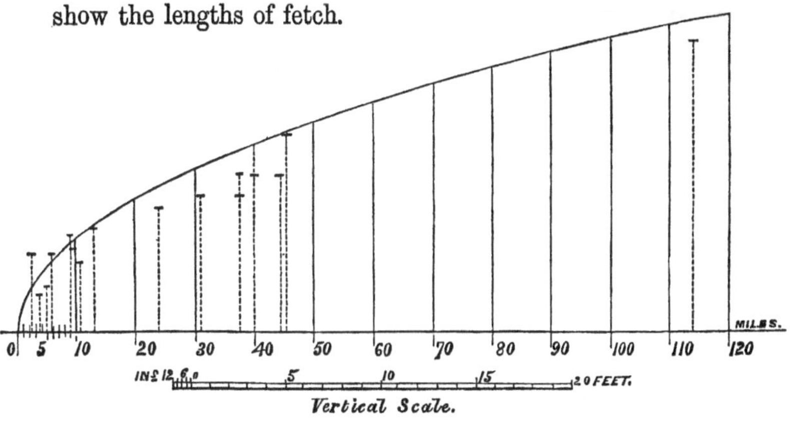

Fig. 1.

Table showing approximate heights of Waves due to lengths of Fetch.

Miles.	Heights.	Miles.	Heights.	Miles.	Heights.	Miles.	Heights.
1 =	3·0 ft.	20 =	7·1 ft.	39 =	9·4 ft.	130 =	17·1 ft.
2 =	3·4	21 =	7·2	40 =	9·5	140 =	17·7
3 =	3·8	22 =	7·4	41 =	9·6	150 =	18·4
4 =	4·1	23 =	7·5	42 =	9·7	160 =	19·0
5 =	4·3	24 =	7·6	43 =	9·8	170 =	19·5
6 =	4·6	25 =	7·8	44 =	9·9	180 =	20·1
7 =	4·8	26 =	7·9	45 =	10·0	190 =	20·7
8 =	5·0	27 =	8·0	46 =	10·2	200 =	21·2
9 =	5·3	28 =	8·1	47 =	10·3	210 =	21·7
10 =	5·6	29 =	8·3	48 =	10·3	220 =	22·2
11 =	5·7	30 =	8·4	49 =	10·5	230 =	22·7
12 =	5·9	31 =	8·5	50 =	10·6	240 =	23·2
13 =	6·0	32 =	8·6	60 =	11·6	250 =	23·7
14 =	6·2	33 =	8·8	70 =	12·5	260 =	24·2
15 =	6·3	34 =	8·8	80 =	13·4	270 =	24·6
16 =	6·5	35 =	8·9	90 =	14·2	280 =	25·1
17 =	6·7	36 =	9·0	100 =	15·0	290 =	25·5
18 =	6·8	37 =	9·2	110 =	15·7	300 =	26·0
19 =	7·0	38 =	9·3	120 =	16·4		

3. *Greatest Height of Waves in the Ocean.*—It is obvious that there must be a limit beyond which the waves will not continue to increase in height, however long may be the line of exposure. The maximum height in the Atlantic, as carefully measured by the late Dr. Scoresby, was 43 feet. He also gave 559 feet as the width from crest to crest, 16 seconds as the interval of time between each wave, and 32½ miles per hour as their velocity.[1] Greater waves than those have been recorded, and are given by Cialdi; but by what means their heights were ascertained we do not know.

4. *Force of Waves.*—As regards the greatest heights at which waves exert their force, the following facts may be noticed:—At the Bishop Rock Lighthouse a bell hung from a bracket on the balcony was broken from its attachments at the level of 100 feet above high water (Naut. Mag., vol. xxxi. p. 262). At Unst Lighthouse, in Shetland, a wall 5 feet high and 2 feet thick was thrown down and a door was broken open at a height of 195 feet above the sea. At the Fastnet Light (Plate VI.), off Cape Clear, a mass of stone from 2½ to 3 tons was thrown off the top of the cliff at the level of 82 feet above the sea. Mr. Conolly, inspector of the works, states that the sea rises over the top of the rock in what he describes as "a solid dark mass," at a height of 36 to 40 feet above the base of the lighthouse tower, being an elevation of from 118 to 122 feet above the sea, and this sheet of water and spray is of sufficient density to shut out the daylight of the lee window on the third story. On one occasion, when a heavy jet of water struck the tower, some cups were thrown off a small table. During another storm a 60-gallon cask, full of rain-water (weighing about 6 cwt.), was burst from its lashings on the balcony at a height of 150 feet above

[1] Life of Dr. W. Scoresby, by Dr. Scoresby Jackson. London: 1861, p. 324.

the sea. It is important to observe that in this case the body of water which passes over the Fastnet and strikes violently against the tower is largely due to the existence of a formidable chasm in the rock, and is probably limited to a comparatively small portion of the upper surface. On the rocky islet at Bound Skerry Lighthouse, in Shetland, at 62½ feet above low water, there is a beach of blocks huddled together, some of which are 9½ tons weight. At an elevation of 72 feet, a block of 5½ tons was torn out of its position *in situ*, and there is no doubt that another of 13½ tons was in like manner *quarried out of the rock* by the waves, at the level of 74 feet above the sea. But the greatest force which has been known to be exerted by the waves was at Wick breakwater, where, on one occasion, a monolithic mass of concrete, weighing 1350 tons, was moved bodily from its position in the work, and on a later occasion a mass of no less than 2600 tons was displaced and moved inwards in a similar manner; and in both of these cases the foundations on which the masses rested were not in the least disturbed.

5. *Numerical Values of the Force of the Waves by the Marine Dynamometer.*—In addition to the facts which have been stated, numerical results have also been obtained by means of the marine dynamometer, an instrument which I designed in 1842. As there is no contest to which the old proverb "*fas est ab hoste doceri*" is more applicable than in resisting the surge of the ocean, it may be proper to give such a description of this simple, self-registering instrument as will enable anyone to have it made.

D E F D is a cast-iron cylinder, which is firmly bolted at the protecting flanges C to the rock where the observations are to be made. This cylinder has a circular flange at D. L is a door which is opened when the observation is to be read off. A is a circular disc on which the waves impinge. Fastened to the disc are four guide rods B, which pass through a circular plate *c* (which is screwed down to the flange D), and also through holes in the bottom of the cylinder E F. Within the cylinder there is attached to the

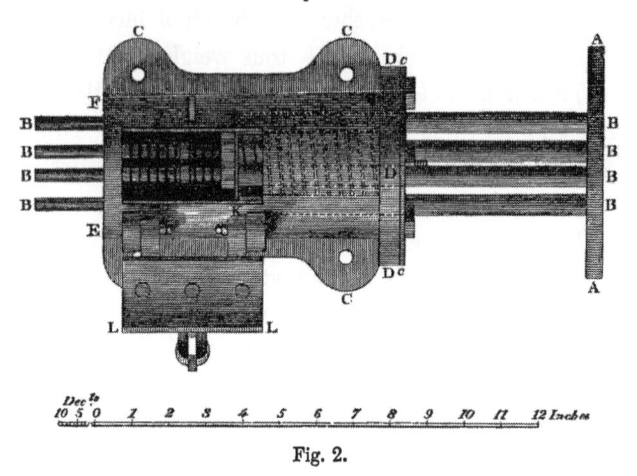

Fig. 2.

plate *c* a very strong steel spring, to the other or free end of which is fastened the small circular plate K, which again is secured to the guide rods B. There are also rings of leather T, which slide on the guide rods, and serve as indices for registering how far the rods have been pushed through the holes in the bottom E F, or, in other words, how far the spring has been drawn out by the action of the waves against the disc A. The following formula will be found convenient in the graduation of the instrument:—

W = weight stated in tons, which is found by experiment to produce a given amount of yielding of the spring. D = number of inches yielded by the spring with weight W. a = area of the disc in square feet, d = the length in inches on the proposed scale, corresponding to a force of one ton per square foot acting on the disc.

$$d = \frac{\mathrm{D}\,a}{\mathrm{W}}$$

The different discs employed in the observations referred to, were from 3 to 9 inches diameter, but generally 6 inches, and the strength of the springs varied from about 10 lbs. to about 50 lbs. for every $\frac{1}{8}$ inch of elongation, and the instruments varied in length from 14 inches to 3 feet. Their respective indications were afterwards reduced to a value per square foot. The instrument was generally placed so as to be immersed at about $\frac{3}{4}$ths tide, and in such situations as would afford a considerable depth of water.

Forces indicated by the Dynamometer.—With instruments of the kind shown in Fig. 2, the following series of observations, commencing in 1843, was made on the Atlantic, at the Skerryvore, and neighbouring rocks, lying off the island of Tyree, Argyllshire; and in 1844 a series of similar observations was begun on the German Ocean, at the Bell Rock.

Referring for more full information to the tables of observations which are given in the Edinburgh Transactions, vol. xvi., it will be sufficient here to state generally the results obtained; only premising that the values refer to areas of limited extent, and are not to be held as simultaneously applicable to large surfaces of masonry.

Relative Force of Summer and Winter Gales.—In the *Atlantic Ocean*, at the Skerryvore Rocks, and at the neighbouring island of Tyree, the average of the results that were registered for five of the summer months during the years 1843 and 1844 was 611 lbs. per square foot = 0·27 ton. The average results for six of the winter months (1843 and

1844) was 2086 lbs. = 0·93 ton per square foot, or more than *three times as great as in the summer months.*

Greatest recorded Forces in the Atlantic and German Oceans.—The *greatest result* obtained at Skerryvore was during the heavy westerly gale of 29th March 1845, when a force of 6083 lbs., or nearly three tons, per square foot, on the surface exposed was registered. The next highest was 5323 lbs.

In the *German Ocean*, the greatest result obtained at the Bell Rock, on the surface exposed, was at the rate of 3013 lbs. per square foot. But subsequent and much more extended observations at Dunbar, in the county of East Lothian, gave 3½ tons; while, at the harbour works of Buckie, on the coast of Banffshire, the highest result of observations extending over a period of several years was three tons per square foot.

In order to show the agreement between the indications of dynamometers having springs of different strength and discs of different area, the following results are given:—

The means of nine observations, which were made with two different instruments, are 433 *lbs. and* 415 *lbs. respectively.*

The means of eighteen observations, which were made with three different instruments, are 1247 *lbs.,* 1183 *lbs., and* 1000 *lbs. respectively.*

6. *Force of the Waves at different Levels above the Sea, as ascertained by the Marine Dynamometer.*—From observations made at Dunbar harbour with dynamometers (Figs. 4, 5) sunk in the masonry with their discs flush with the surface of a curved sea wall, it appeared that while the greatest force was exerted at the level of high water, the forces quickly decreased above and below that level, as shown by *bb, cc, dd,* Fig. 3. Owing to an uncertainty as to the readings of some of the instruments, which was not discovered at the

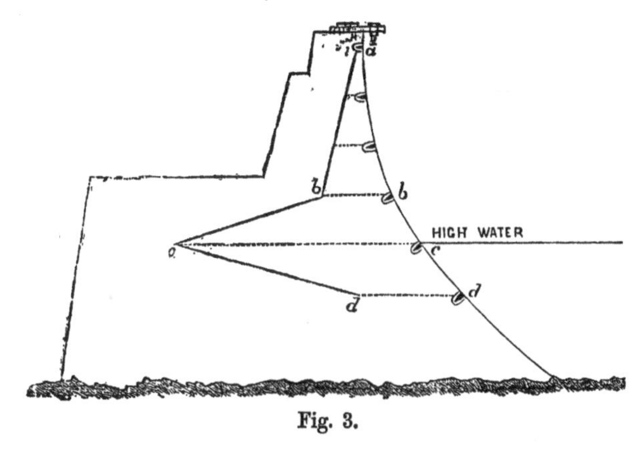

Fig. 3.

time, the results must be regarded as only approximate. Two dynamometers (*a*, Fig. 3), similar to that shown in Fig. 2, were also fixed to a piece of wood projecting over the cope of the parapet of the wall, with their discs pointing downwards over the edge of the cope, so as to ascertain the upward force of the ascending body of water and spray. The *maximum vertical force* recorded at the cope of the wall (23 feet above the sea at the time the observation was made) was upwards of one ton (2352 lbs.) per superficial foot, while

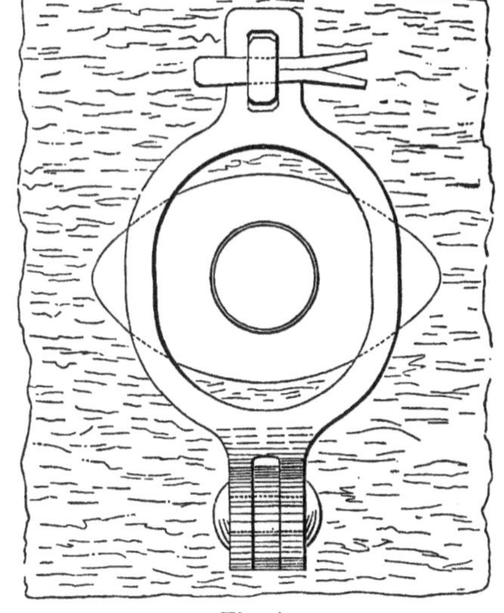

Fig. 4.

the greatest *horizontal* force which was at *any* time recorded by the highest of the *flush* dynamometers, which was 18 inches lower, never exceeded 28 lbs. The *vertical*

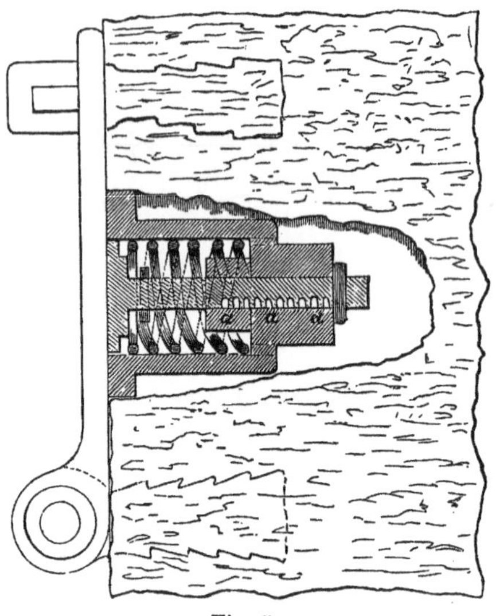

Fig. 5.

force tending to raise the cope of this sea-wall, if projecting, was therefore at least 84 *times greater than the horizontal force tending to thrust it inwards.*

7. *Horizontal Force of the Back Draught of the Wave.*—Dynamometers having their discs pointed landwards were fixed to a pile placed immediately outside of the Dunbar wall, while others were fixed to the same pile with their discs pointed seawards. The force of recoil was found to be one ton per square foot, while the direct force of the waves before they had reached the wall was only 7 cwts. per square foot, or *the force of recoil with this form of wall was three times that of the direct force*—a result which was doubtless due to the con-

centration of energy, and was produced by the resistance of the masonry.

8. *Mr. Scott Russell's Law of the Relation of Height of Waves to Depth of Water.*—Mr. Scott Russell states that if waves be propagated in a channel whose depth diminishes uniformly, the waves will break when their height becomes equal to the depth of the water. The depth for some distance round any rock on which a lighthouse is placed becomes, therefore, the ruling element which determines the height of the impinging waves on the masonry of the tower, whatever may be their height farther out in the ocean. But the law assigned by Mr. Russell does not appear to hold true of all waves. At Scarborough I measured waves 5 feet 3 inches high, which broke in 10 feet 3 inches water, and others 6 feet 6 inches high, which broke in 13 feet 8 inches water, so that when h = height in feet of waves of this class, as measured from hollow to crest, and D depth of water in feet below the mean level—

$$h = \frac{D}{2}$$

At Wick breakwater waves were seen to break in water of the same depth as their height. The height of these waves above mean level was about $\frac{2}{3}$ of their height, the hollow below mean level being about $\frac{1}{3}$.

The late Dr. Macquorn Rankine has shown, on theoretical grounds, that the mean level is not equidistant between the crest and the trough.

Let L be the length of a wave, H the height from trough to crest; then, diameter of rolling circle $= \frac{L}{3 \cdot 1416}$; radius of orbit of particle $= \frac{H}{2}$; and elevation of *middle level* of wave above still water $= \frac{3 \cdot 1416 H^2}{4L} = \cdot 7854 \frac{H^2}{L}$

Consequently—

$$\text{Crest above still water} = \frac{H}{2} + \cdot 7854 \frac{H^2}{L}$$

$$\text{Trough below still water} = \frac{H}{2} - \cdot 7854 \frac{H^2}{L}$$

These formulæ are exact for water of considerable depth, and only approximate for shallow water.

LIGHTHOUSE TOWERS.

9. *Remarkable Lighthouse Towers.*—In giving a description of some of the most remarkable towers we shall omit all the earlier works, the history of which is more or less shrouded in obscurity, and is therefore more curious than useful, and shall restrict ourselves principally to those works only, which are exposed to heavy seas, and which furnish us with facts from which practical lessons can be deduced. Plate No. I. contains drawings of those towers on the same scale, and placed at their relative levels above the sea.

Winstanley's Eddystone Light.—The Eddystone rocks, lying about 14 miles off the harbour of Plymouth, are fully exposed to the south-western seas. In 1696 Mr. Henry Winstanley, of Littlebury, Essex, undertook to erect a tower on the Eddystone. The work was begun in that year, and completed in four seasons. Twelve large iron stanchions were fixed during the first year. The erection of a solid cylinder of masonry 12 feet high and 14 feet diameter occupied the second year. This column was in the third year (1698) increased in diameter to 16 feet, and finished by raising it to the height of 80 feet, when the light was exhibited. In the next year (1699), in consequence of some damage that occurred to the building during the winter, the tower was encircled by an outer ring of masonry of 4 feet in thickness, and made solid from the foundation to nearly 20 feet above the rock. The upper part of the tower was taken down, and besides being enlarged proportionally throughout, was raised 40 feet higher, making the total height 120 feet, the whole work being completed in the year 1700. On the 26th November 1703, during the greatest British storm of which we have any record, the structure disappeared, and, as is well known, Winstanley, who had gone to make some repairs, perished with his workmen.

Rudyerd's Eddystone Light.—In 1706 a lease of 99 years was given by the Trinity House of London to a Captain Lovet, who was to erect another tower on this rock. Captain Lovet made a strange selection of an engineer—J. Rudyerd, a silk mercer in Ludgate Hill—who, however, had the good sense to engage experienced shipwrights to assist him. The work was commenced in July 1706, and completed in 1709. The form which he adopted was the frustum of a cone, the height from the base to the ball on the top of the lantern being 92 feet. The rock was cut out in level steps. The lower portion of the tower consisted of oak timber, carefully bolted together and to the rock by an ingenious system of iron bars and jag bolts. Above this mass of carpentry were placed courses of stone cramped together, and connected with the timber-work and with the rock. The upper portion of the cone consisted wholly of timber. The entrance-door was 8 feet above the foundation, the storeroom floor 27 feet above it, and higher up were four rooms with lantern and balcony. This lighthouse stood for the long period of 46 years, and was at last destroyed by fire in December 1755.

Smeaton's Eddystone Tower.—This celebrated work (Fig. 6), which was constructed entirely of stone, was commenced in August 1756 by cutting dovetailed holes in the rock, and in the

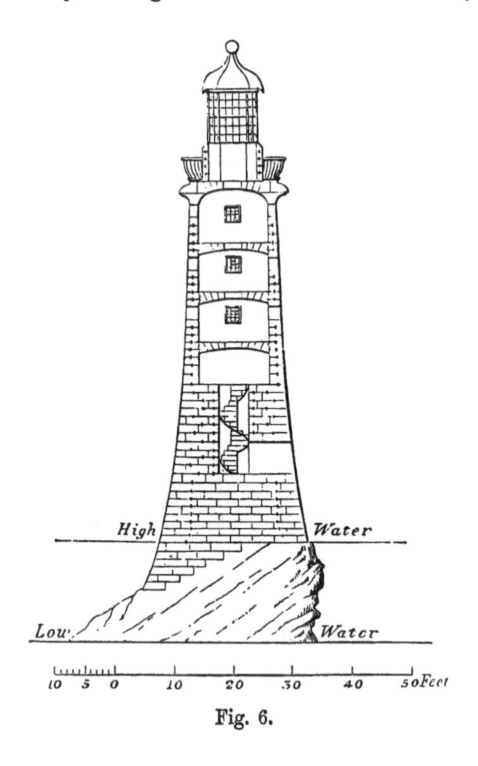

Fig. 6.

second year (1757) the masonry was brought up to the top of the ninth course. In 1758 the tower was nearly ready for the light, but was stopped in consequence of some difficulties connected with the Act of Parliament. In August 1759 the whole of the masonry was finished. Smeaton was the first lighthouse engineer who adopted dovetailed joints for the stones, which averaged one ton in weight. In Plate II. will be found drawings of the ingenious method which he employed for keeping them together during construction. He adopted (Fig. 7) a vaulted form for the floors of the different apartments, the voussoirs of the arch being connected

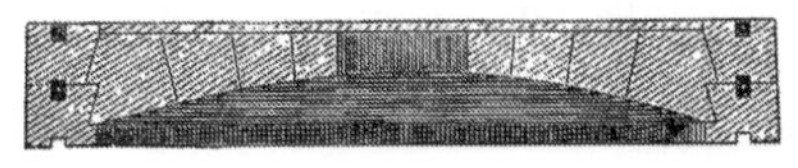

Fig. 7.

together so as to form a continuous mass, in addition to which he inserted metallic hoops (shown black in diagram) to prevent lateral thrust or tendency to spread outwards.[1] The damage which has recently led the Trinity House to the resolution of removing the present tower and to build a new one will be afterwards referred to.

Bell Rock.—The Inchcape or Bell Rock lies 12 miles off the coast of Forfarshire, and is fully exposed to the waves of the German Ocean. The rock is of large dimensions, the higher part being 427 feet in length and 230 feet in breadth, and though the sandstone of which it consists is of a somewhat indurated character, it lies in ledges which are easily dislodged by the sea; and since the erection of the tower, portions of the rock near the base have been frequently detached, leaving voids which have from time to time been made up with cement or concrete. The place on which the tower is erected is 16 feet below high-water spring tides, or not much above the level of low-water springs, while during neap-tides hardly any part of this reef is visible at low water. Captain Brodie proposed an iron pillar light, presumably the first proposition of the kind, and in order to prove its stability he erected two structures of timber at different dates, but they were both removed by the waves. In 1799 the late Mr. Robert Stevenson, engineer to the Northern Lighthouse Board, also prepared, before visiting the rock, a plan for

[1] Smeaton's Narrative of the Building of the Eddystone Lighthouse; London, 1791.

an iron pillar lighthouse. In 1800 Mr. Stevenson made his first landing on the reef, when, as he expressed it, "I had no sooner set foot upon the rock than I laid aside

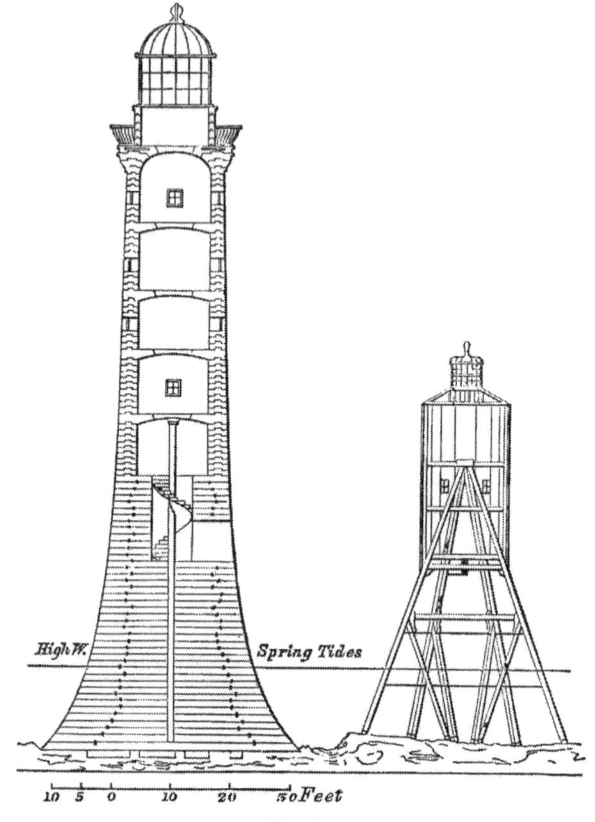

Fig. 8.

all idea of a pillar-formed structure, fully convinced that a building on similar principles with the Eddystone would be found practicable." In his design (Fig. 8) he deviated in several respects from Smeaton's tower; increasing the thickness of the walls, and raising the height to 100 feet instead of 68 feet, and the level of the solid to 21 feet above high water instead of 11 feet, while the base was made 42 feet instead of 26 feet. Instead of employing vaulted floors, as in the Eddystone, he converted them into effective bonds, which tie the walls together (Fig. 9). The lintel stones of the floors form parts of the outward walls, and are feathered and grooved as in carpentry, besides having dovetailed joggles across the joints where they form part of the walls. He also used a temporary beacon or barrack (Fig. 8, and Plate IV.), which was erected on the rock for the engineer and his workmen to live in, during the building of the tower. The first Act of Parliament was applied for in 1802, but was not obtained owing to financial difficulties. As the Bell Rock was *scarcely dry at low water*, while *the Eddystone was scarcely covered at high water*, the Commissioners, in order to fortify Mr. Stevenson's views, consulted Mr. Telford; and before going to Parliament for the second time they also, on Mr. Stevenson's suggestion, obtained the support of Mr. Rennie to the scheme, with whom Mr. Stevenson could afterwards advise in case of emergency during the progress of the work. The second bill was passed in 1806, and the works were begun in 1807. In 1808 the excavation was finished, and the masonry brought up to the level of the rock. In 1809 the work was carried up to 17 feet above high water. In 1810 the whole tower was finished, and the light was exhibited in 1811. The total weight of the tower is 2076 tons.[1]

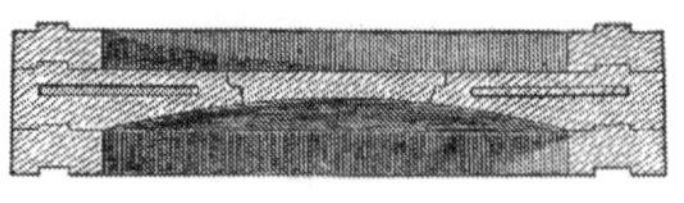

Fig. 9.

Skerryvore Lighthouse.—The Skerryvore rocks in Arygllshire lie 12 miles off the island of Tyree, which is the nearest

[1] Account of the Bell Rock Lighthouse, by R. Stevenson. Edin. 1824.

land. The main rock is of considerable extent, and being gneiss, is of great hardness. The works, which were designed and carried out by the late Mr. Alan Stevenson, were com-

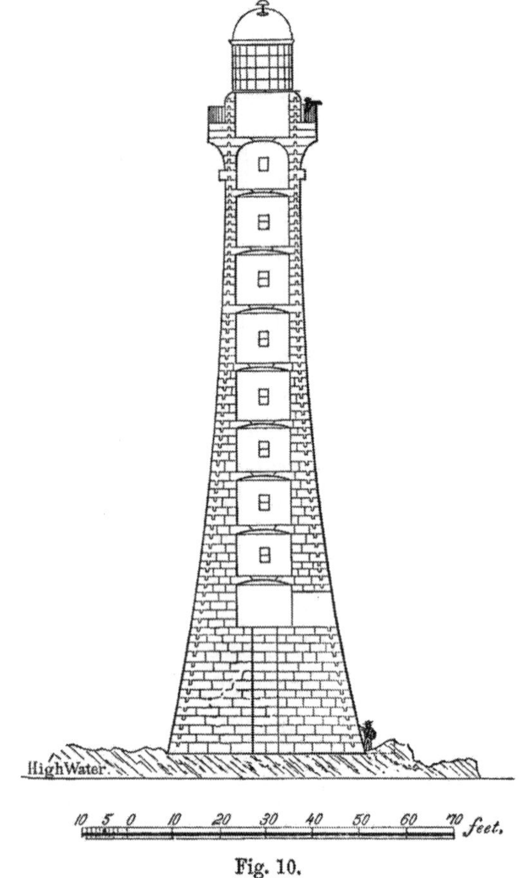

Fig. 10.

menced in 1838 by the erection of the framework of a barrack similar to that used at the Bell Rock, but it was carried away in November of the same year, after which another was erected in a more sheltered position. The works were not finished until 1843. The tower (Fig. 10), which is of granite, is 138 feet high, 42 feet diameter at the base and 16 feet at the top. It contains a mass of stonework of 58,580 cubic feet, or more than double that of the Bell Rock, and not much less than five times that of the Eddystone. The curve adopted for the shaft of the tower was the hyperbolic.[1]

Fastnet Rock.—The Fastnet lies about 4½ miles to the westward of Cape Clear, and is a light of great importance to all oversea vessels making the south coast of Ireland as their landfall. The rock, which is surrounded by deep water in almost every direction, attains a maximum height of 87 feet above high water, the base of the tower being about 70 feet above that level; it is about 340 feet long and 180 feet broad at low water, and on the top about 120 feet long and 64 feet broad. The lighthouse (Plate VI.) is 19 feet diameter at base and 63 feet high, and consists of iron plates weighted with concrete at the bottom. The structure is also fixed to the rock by bolts 2 feet long. The iron casing is lined with brick varying from 2 feet 9 inches thick at bottom to 9 inches at top. The dwelling-houses, which are built close to the lighthouse, are made of plates of cast iron. The sea has excavated some of the softer slates of which the rock consists, forming different chasms, one of which, facing the west, extends from about 36 feet above low water to the top, and varies in breadth from about 1 foot to 13 feet, and penetrates inward from the face of the cliff about 14 feet. The top of this fissure extends to within 10 feet of the tower. This lighthouse has been strengthened since its erection, and the fissure in the rock has been filled up with brickwork set in cement.

[1] Account of the Skerryvore Lighthouse, by Alan Stevenson, LL.B. Edin. 1848.

Bishop Rock.—The Bishop Rock lies off the Scilly Isles. The first lighthouse was an iron pillar structure, which was unfortunately destroyed before it was completed; and in 1852-8 the present stone structure was erected. Both of these designs were by the late Mr. James Walker, and the present tower was carried out under the superintendence of Mr. J. N. Douglass. This lighthouse rises to the height of 100 feet above high water, is 34 feet in diameter at the base and 17 feet at top. The lowest part of foundation of tower is covered about 19 feet at high-water springs. The "solid" is 20 feet above high water, where the walls are 9 feet in thickness, decreasing to 2 feet at the top. The interior is divided into five stories, each 13 feet in diameter. The first stone was laid in July 1852, and the last in August 1857. Mr. Douglass states that "during a storm in the winter of 1874-75 the heaviest seas experienced since the completion of the lighthouse in 1858 were encountered. On this occasion the tremor of the building was so great as to cause articles to leave the shelves. On his (Mr. Douglass') recommendation, the building has since been strengthened by strong internal vertical and radiating ties of wrought iron screwed to the walls and floors."

Wolf Rock.—This rock lies about midway between Scilly and the Lizard Point, and is a hard porphyry. It is about 170 feet long and 114 feet broad, and is submerged to the depth of about 2 feet, at high water. The first design for this work was in 1823 by Mr. Robert Stevenson. But it was not till 1862 that a lighthouse was commenced from a design by the late Mr. James Walker, under the superintendence of Mr. James N. Douglass. Its height is 116½ feet, 41 feet 8 inches diameter at base, decreasing to 17 feet at the springing of the cavetto course. The "solid," excepting a space for a water-tank, extends to about $39\frac{1}{2}$ feet in height, the stones being laid with off-sets or scarcements in order to break up the sea; but the surface above the solid is smoothly cut. The walls at entrance-door are 7 feet $9\frac{1}{2}$ inches thick and 2 feet 3 inches at top. The shaft of the tower is a concave elliptic frustum of granite, and contains $3296\frac{3}{4}$ tons.[1] It is founded 2 feet above low water.

The Dhu Heartach Lighthouse is 14 nautic miles from the Island of Mull, which is the nearest shore. The rock, which is a hard trap, is 240 feet long, 130 feet broad, with a rounded top rising to 35 feet above high water. The curve which was found best suited for this rock was the parabolic. It was considered, on engineering grounds, that the lines of the parabolic shaft should run continuously into the cavetto without any belt course, which in this particular case, owing to the high level to which the water is projected, would have had to oppose a very considerable force, tending to shake the masonry. The maximum diameter of the tower is 36 feet; the minimum 16 feet; the level of the solid was ultimately fixed at 32 feet above the rock. The total amount of masonry is 3115 tons, of which 1810 are contained in the solid part. The barrack for the workmen (Plate V.) was made of malleable iron bars, with a malleable iron drum placed on the top, in which the men lived. The height of lower floor of the barrack above the rock was 35 feet. This tower, which occupied six years in erection, was designed by Messrs. D. and T. Stevenson, and was carried out under the immediate superintendence of Mr. Alexander Brebner. Owing to the great difficulty of landing, the working seasons were limited to only about $2\frac{1}{2}$ months in the year.[2]

[1] The Wolf Rock Lighthouse. By J. N. Douglass. Min. Civ. Eng. vol. xxx.
[2] The Dhu Heartach Lighthouse. By D. A. Stevenson. Min. Civ. Eng. vol. xlvi.

The Chickens Rock lies one mile off the Calf of Man. The tower is a frustum formed by the revolution of a hyperbola about its asymptote as a vertical axis, surmounted by a belt course and plain cavetto, abacus and parapet. It is 123 feet 4 inches in height; the outside diameter is 42 feet at the base and 16 feet at the plinth. The tower is submerged 5 feet at high water of spring tides, and lies open to the south-westerly seas entering St. George's Channel. The solid is 32½ feet in height, and contains 27,922 cubic feet of masonry, weighing 2050 tons, the whole weight of the tower being 3557 tons. The walls, as they rise from the solid, are 9 feet 3 inches in thickness, decreasing to 2 feet 3 inches below the belt course. The work, which was designed by Messrs. D. and T. Stevenson, was begun in April 1869, and completed in 1874.

Great Basses Lighthouse.—The Great Basses Reef lies about 80 miles to the eastward of Point de Galle, and 6 miles from the nearest land. The lighthouse was designed by Mr. James N. Douglass, and is placed on a hard red sandstone rock, measuring 220 feet in length, 75 feet in breadth, and rising 6 feet above the mean sea level. The extreme range of the tide is about 3 feet, and the foundation of the tower is 2 feet above high water. The tower consists of a cylindrical base 30 feet in height, 32 feet in diameter, surmounted by a tower 67 feet 5 inches in height, 23 feet in diameter at base, and 17 feet at the springing of the cavetto, the thickness of walls being 5 feet at the base and 2 feet at the top. The cylindrical base is solid for 11½ feet. The tower and base contain 37,365 cubic feet of granite masonry, weighing about 2768 tons. The work was begun on 5th March 1870, and the light was exhibited three years afterwards.[1]

[1] The Great Basses. By W. Douglass. Min. Civ. Eng., vol. xxxviii.

Fig. 11 shows the Cordouan Tower, in France, begun in

Fig. 11.

the time of Louis XIV., which is given more as a specimen

of early lighthouse architecture than of a tower which is much exposed to the sea.

There are several other lighthouses in situations more or less exposed, such as the Horsburgh, near Singapore, designed and executed by Mr. J. T. Thomson; that on the Alguada Reef, which, though somewhat higher, is, in other respects, a *replica* of the Skerryvore, and was successfully carried out by Captain Fraser; the Prongs in India, by Mr. Ormiston; the lights in the Red Sea by Mr. Parkes; the South Rock and Haulbowline in Ireland; Craighill Channel (Plate VII.) and Spectacle Reef in America (Plate VIII.), which are exposed to the action of moving floes of ice. Those which have been described in detail are the most remarkable, and some of them will be referred to in illustration of the principles of construction, which will now be explained.

General Principles which regulate the Design of Lighthouse Towers exposed to the Waves.

10. We cannot do better, as a preliminary to the subject of Lighthouse construction, than quote the following remarks by the late Mr. Alan Stevenson:[1]—

"A primary inquiry as to towers in an exposed situation, is the question, Whether their stability should depend upon their strength or their weight; or, in other words, on their cohesion or their inertia? In preferring weight to strength we more closely follow the course pointed out by the analogy of nature; and this must not be regarded as a mere notional advantage, for the more close the analogy between nature and our works, the less

---

[1] Account of Skerryvore Lighthouse. By Alan Stevenson, LL.B. Edinburgh, 1848, page 49.

difficulty we shall experience in passing from nature to art, and the more directly will our observations on natural phenomena bear upon the artificial project. If, for example, we make a series of observations on the force of the sea as exerted on masses of rock, and endeavour to draw from those observations some conclusions as to the amount and direction of that force, as exhibited by the masses of rock which resist it successfully, and the forms which those masses assume, we shall pass naturally to the determination of the mass and form of a building which may be capable of opposing similar forces, because we conclude, with some reason, that the mass and form of the natural rock are exponents of the amount and direction of the forces they have so long continued to resist. It will readily be perceived that we are in a very different and less advantageous position when we attempt, from such observations of natural phenomena in which weight is solely concerned, to deduce the strength of an artificial fabric capable of resisting the same forces; for we must at once pass from one category to another, and endeavour to determine the strength of a comparatively light object which shall be able to sustain the same shock, which we know, by direct experience, may be resisted by a given weight. Another very obvious reason why we should prefer mass and weight to strength, as a source of stability, is, that the effect of mere inertia is constant and unchangeable in its nature; while the strength which results, even from the most judiciously-disposed and well-executed fixtures of a comparatively light fabric, is constantly subject to be impaired by the loosening of such fixtures, occasioned by the almost incessant tremor to which structures of this kind must be subject, from the beating of the waves. Mass, therefore, seems to be a source of stability, the effect of which is at once apprehended by the mind, as more in harmony with the conservative principles of nature, and unquestionably less liable to be deteriorated than the strength which depends upon the careful proportion and adjustment of parts.

* * * * *

"Of any solid, viewed as monolithic, it may be said that its ultimate stability, by which I would understand its resistance to the final effort which overturns it, will greatly depend upon its centre of gravity being placed as low as possible; and the general sectional form which this notion of stability indicates is that of a triangle. This figure revolving on its vertical axis, must, of course, generate a cone as the solid, which has its centre of gravity most advantageously placed, while its rounded contour would oppose the least resistance which is attainable in every direction. Whether, therefore, we make strength or weight the source of stability, the conic frustum seems, abstractly speaking, the most advantageous form for a high tower. But there are various considerations which concur to modify this general conclusion, and, in practice, to render the conical form less eligible than might at first be imagined. Of these considerations, the most prominent theoretically, although, I must confess, not the most influential in guiding our practice, is, that the base of the cone must in many cases meet the foundation on which the tower is to stand, in such a manner as to form an angular space in which the waves may break with violence. The second objection is more considerable in practice, and is founded on the disadvantageous arrangement of the materials, which would take place in a conic frustum carried to the great height which, in order to render them useful as sea marks, lighthouse towers must generally attain. Towards its top the tower cannot be assaulted with so great a force as at the base, or rather, its top is entirely above the shock of heavy waves; and as the diameter of the conoidal solid should be proportioned to the intensity of the shock which it must resist, it follows that, if the base be constructed as a frustum of a given cone, the top part ought to be formed of successive frusta of other cones, gradually less slanting than that of the base. But it is obvious that the union of frusta of different cones, independently of the objection which might be urged against the sudden change of direction at their junction, as affording the waves a point for advantageous assault, would form a figure of inharmonious and unpleasing contour, circumstances which necessarily lead to the adoption of a curve osculating the outline of the successive frusta composing the tower; and hence, we can hardly doubt, has really arisen in the mind of Smeaton the beautiful form which his genius invented for the lighthouse tower of the Eddystone, and which subsequent engineers have contented themselves to copy, as the general outline which meets all the conditions of the problem which they have to solve. And here I cannot help observing, as an interesting and by no means unusual psychological fact, that men sometimes appear to be conducted to a right conclusion by an erroneous train of reasoning: and such, from his 'Narrative,' we are led to believe must have been the case with Smeaton in his own conception of the form most suitable for his great work. In the 'Narrative' (§ 81), he seems to imply that the trunk of an oak was the counterpart or antitype of that form which his (§ 246) 'feelings, rather than calculations' led him to prefer. Now, there is no analogy between the case of the tree and that of the lighthouse, the tree being assaulted at the top, and the lighthouse at the base; and although Smeaton goes on, in the course of the paragraph above alluded to, to suppose the branches to be cut off, and water to wash round the base of the oak, it is to be feared the analogy is not thereby strengthened; as the materials composing the tree and the tower are so different, that it is impossible to imagine that the same opposing forces can be resisted by similar properties in both. It is obvious, indeed, that Smeaton has unconsciously contrived to obscure his own clear conceptions in his attempt to connect them with a fancied natural analogy between a tree which is shaken by the wind acting on its bushy top, and which resists its enemy by the strength of its fibrous texture and wide-spreading ligamentous roots, and a tower of masonry whose weight and friction alone enable it to meet the assault of the waves which wash round its base; and it is very singular that, throughout his reasonings on this subject, he does not appear to have regarded those properties of the tree which he has most fitly characterised as 'its elasticity,' and the 'coherence of its parts.' * * * *

"In a word, then, the sum of our knowledge appears to be contained in this proposition: That, as the ultimate stability of a sea tower, viewed as a monolithic mass, depends, *cœteris paribus*, on the lowness of its centre of gravity, the general notion of its form is that of a cone; but that, as the forces to which its several horizontal sections are opposed decrease towards its top in a rapid ratio, the solid should be generated by the revolution of some curve line convex to the axis of the tower, and gradually approaching to parallelism with it. And this is, in fact, a general description of the Eddystone tower devised by Smeaton."

Though the views which have been so well expressed, and the general conclusions at which Mr. Alan Stevenson arrived, are unquestionably correct, some modifications and extensions of these seem, nevertheless, to be required in order to give a complete view of the subject. The following rules may be regarded as of general application to the design of all towers of masonry in exposed situations:—

1. They should have a low centre of gravity, and sufficient mass to prevent their being upset by the waves.

2. They should be throughout circular in the horizontal plane, and either straight or continuously curved in the vertical plane, so as to present no abrupt change of outline which would check the free ascent of the rising waves, or the free descent of the falling waves, or the free vent of those passing round the tower. All external scarcements in the vertical plane, or polygonal outline in the horizontal plane, are therefore objectionable.

3. Their height, *cæteris paribus,* should be determined by the distance at which the light requires to be seen by the sailor. The rule for determining this height will be afterwards given.

4. Where the rock is soft, or consists of ledges which are easily torn up, the tower should spring from the foundation-course at a low angle with the surface of the rock, so as to

prevent its being broken up by reaction of the waves from the building; or, in other words, the tower must have a curved profile, as shown in Fig. 12. But especial care should be taken to sink the foundation-courses below the surface of the rock, as the superincumbent weight decreases with the sine of the angle of inclination of the wall. If the rock overhangs, owing to the wearing action of the waves, the tower should, if possible, be built at a distance from the place where this dangerous action is in progress.

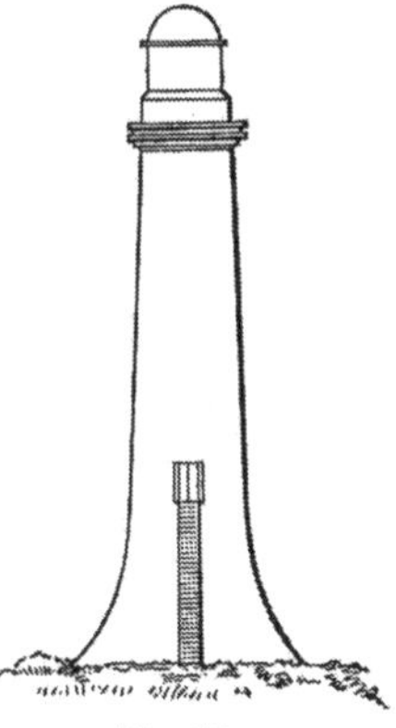

Fig. 12.

5. Where the rock is hard and of ample area, the tower may be of such a curved form as will best suit the economic arrangement of the materials, so as to avoid an unnecessary thickness of the upper walls.

6. Where the rock is hard, but of small dimensions, the diameter above the base should not be suddenly reduced by adopting a curved profile, but a conical outline should be adopted (as in Fig. 13); and if the rock be hard, but of yet smaller dimensions, a cylindric form should be adopted to thicken the walls (Fig. 14), and to increase the friction, which is directly proportional to the weight, the tower should be raised higher than if the diameter had been greater. In all cases where the rock is small the thickness of the walls should be decreased by steps or scarcements *internally*, as in Fig. 14, but *never externally*.

7. The level of the top of the solid part of the tower, and the thickness of the walls above it, should, in different

towers having the same exposure, be determined in each case by the level and configuration of the rock and of the bottom of the sea. Unfortunately, in the present state of our knowledge, no rule can be given for dealing with such cases. But facts, illustrative of the great influence of the configuration of the rock itself on the action of the waves, are given in a subsequent section.

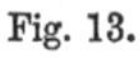

Fig. 13.

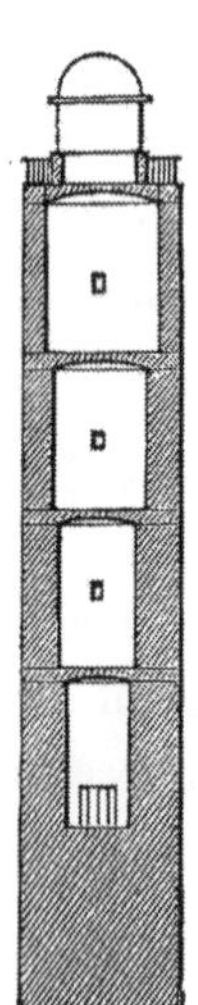

Fig. 14.

8. The best position for the tower is not necessarily the highest part of the rock. It should, in each case, be selected so as to secure the greatest protection in the direction of the maximum fetch and deepest water near the reef.

9. The tower should not, if possible, be erected across any chasm which divides the rock, or in the direction of any gully which projects into it, especially if it be of converging form which would concentrate wave action.

10. No permanent fixture of the tower to the rock is

required for increasing the stability of the structure. The foundation-course (unless where a curved profile is adopted) becomes, indeed, in the end the most stable of all, because it has the greatest weight above it, to keep it in its place.

11. The stones should, however, be sufficiently connected together, and fixed to the rock, in order to prevent their being washed away during the construction of the work, when they have no superincumbent weight to keep them in their beds.

12. The tower should rest on a truly level base, or on level steps cut in the rock.

13. The pressure of all the materials within the tower should act vertically, so as not to produce a resolved force acting laterally as an outward thrust.

14. The tower should be of such height and diameter, with walls of such thickness, as to prevent the masonry being disturbed by the impact of the waves.

15. The entrance door should be placed on that side of the tower where the length of fetch is shortest, or where, from the configuration of the reef and the depth of water, the force of the waves is least. This was determined at the Bell Rock by the distribution of the *fucus* which grew on the lower parts of the tower during the winter, the vegetation being least where the waves were heaviest.

16. The materials should be of the highest specific gravity that can be readily obtained, and, in some special cases, lead, or dovetailed blocks of cast-iron set in cement, might perhaps be employed.

Application of Foregoing Rules to the Construction of Towers.

11. There are no better examples of what to avoid and what to imitate in lighthouse construction than are to be found in the towers which were erected on the Eddystone rock. The first of these (Winstanley's), both in general design and details, must be placed in the *vetanda* of maritime engineering, while Rudyerd's and Smeaton's contain between them what seem to be the best models.

Winstanley's Tower.—This tower is so consistently bad in all its parts that it is unnecessary to do more than simply state its defects.

1. He failed to take full advantage of the available base which the rock, small as it is, afforded. This error he no doubt did his best to rectify during the subsequent years in which the work was in progress.

2. Instead of being circular in the horizontal plane, it was polygonal.

3. In his blind devotion to ornamentation he violated the principle of uniformity of external surface.

4. The only redeeming point in the structure was that he did not, in the end, throw away the diameter of the foundation by adopting a curved vertical outline.

Rudyerd's Tower.—In whatever direction we look, this tower, conspicuous for its simplicity of general design and careful adaptation of the different parts, will be found to fulfil the conditions of almost every axiom we have mentioned.

1. It was erected on a level base; or at least the slanting ledge was cut out in level steps, so far as the hard nature of the stone would allow.

2. It was circular in the horizontal plane, instead of polygonal, as in Winstanley's tower.

3. The carpentry was most judiciously constructed; but, as it had not sufficient weight in itself, it required to be bolted to the rock, and the bolting was of an ingenious and effective kind.

4. He did not, however, depend wholly upon strength of fixtures, but upon *weight*, as appears from his placing courses of masonry above the lower portion of the timber-work.

5. He did not needlessly throw away the diameter afforded by the size of the rock, but adopted the conical form, sloping from the outside at the bottom to the size required for the light-room accommodation at the top.

6. He preserved a uniform external surface by avoiding all outside projections and ornamentation; even the window shutters were made to close *flush* with the outer face.

7. There were no arches in any part of the building tending to produce an outward lateral thrust.

The only defect was in the want of weight and of durability of some of the materials, and their liability to catch fire. It may, perhaps, be said that he should have used as little wood and as much stone as he could, because it was heavier; but it must be recollected that he acted in the matter as a contractor, and perhaps thought himself justified in employing only as much weight as he considered sufficient; and in this he judged correctly, for the building withstood all the storms of forty-six years, and yielded at last, not to the action of the waves, but to an accidental fire, which destroyed the work. There is probably no instance in the records of the profession of such a wholly untaught engineer being found able at once to grapple with the difficulties of so novel a problem and eventually to produce a true masterpiece of construction. It is further remarkable that, after having achieved his difficult work, he seems suddenly to have retired either to his mercer's shop or into private life, instead of, like his great successor, becoming the leading engineer of the day. So far as is known, the Eddystone was his first and last work.

Smeaton's Tower.—This tower cannot be said to have been in any respect more successful as regards resisting the waves than Rudyerd's, which was the earlier of the two. The success of Smeaton's Lighthouse is no doubt its best vindication; but it should not, as I have elsewhere said, be regarded as a safe model for imitation at all places which are exposed to a very heavy sea. For rocks of such small area it would surely be safer to adopt a conical form like that of Rudyerd's. Having once secured a certain diameter for the foundation on the rock, that diameter should not, in my opinion, have been suddenly thrown away higher up by adopting a sharply-curved profile. The case of an oak tree resisting the wind has been already satisfactorily shown to be in no way analogous to that of a tower exposed to the surf; and greatly as this curve has been admired, it ought not to be followed in all cases. The advantages which it is supposed to possess seem, in such a situation as the Eddystone, to be more fanciful than real, and certainly not for a moment to be compared to the loss of diameter resulting from its adoption. The arched floors were also a source of weakness, which the introduction of the iron hoop was intended to counteract.

The rock at the Eddystone, though of great hardness, was even in Smeaton's time considerably hollowed out underneath by the grinding action of the waves. In 1818 Mr.

R. Stevenson, on visiting the Eddystone, made the following remarks as to the wasting of the lower part of the rock:—"I conclude that, when the sea runs high, there is danger of this house being upset after a lapse of time, when the sea and shingle have wrought away the rock to a greater extent. Nothing preserves this highly important building but the hardness of the rock and the dip of the strata, but for how long a period this may remain no one can pretend to say."[1]

The fears thus expressed as to the wasting of the rock, and my own convictions in 1864, and also those of Captain Fraser[2] as to the insufficiency of the tower itself for exposed situations, have since been verified. Mr. Douglass[3] in 1878 said:—"For several years the safety of the Eddystone Lighthouse has been a matter of anxiety and watchful care to the Corporation of the Trinity House, owing to the great tremor of the building with each wave-stroke during heavy storms from the westward, more especially when from west-south-west. The joints of the masonry have frequently yielded to the heavy strains imposed on them, and the sea water has been driven through them to the interior of the building. The upper part of the structure has been strengthened on two occasions—viz. in 1839 and again in 1865—with strong internal wrought-iron ties, extending from the lantern floor downwards to the solid portion of the tower. On the last occasion it was found that the chief mischief was caused by the upward stroke of the heavy seas acting on the projecting cornice under the lantern gallery, which lifted the portion of the building above this level. After reducing the projection of the cornice about 5 inches, and well fasten-

[1] Life of Robert Stevenson. By David Stevenson, Edinburgh. 1878. P. 47.

[2] Design and Construction of Harbours. Edinburgh, 1864.

[3] Min. of Pro. of the Inst. of Civil Engineers. Vol. liii.

ing the stones together with through bolts, no further serious leakage has occurred at this part."

Mr. Douglass says further as to the action of the waves on the rock itself:—"The portion of the gneiss rock on which the tower is founded has been seriously shaken by the incessant heavy sea-strokes on the tower, and the rock is considerably undermined at its base." As already stated, the Trinity House have resolved on the removal of Smeaton's tower, and the erection of a new lighthouse from the design of Mr. Douglass, which is now being carried out.

12. *Modifying influence of the configuration of rocks on breaking waves.*—In November 1817, the waves of the German Ocean overturned, just after it had been finished, the upper part of the Carr Rock Beacon, a column of freestone 36 feet high, and 17 feet diameter at the base, which was the largest diameter that could be got on the reef. A curved profile was adopted, but this seems to have been warranted by the unreliable character of the sandstone on which it was built. The diameter at the level of high water was 11 feet 6 inches, and at the plane of fracture 12 feet 9 inches. When the history of this beacon is contrasted with the records which have been preserved of the different structures that were erected on the Eddystone by Winstanley, there is every reason to suspect that the configuration of that rock must have the effect of modifying or diverting the full force of the waves. As the additions which Winstanley made to his original tower were in many respects anything but improvements, it may be questioned whether its fate was not accelerated rather than delayed by those alterations. It stood, however, until the fifth winter, when, during the greatest storm ever experienced in this country, it was swept away. From the records of that tempest I think it may fairly be questioned whether Winstanley's work was really knocked down by the sea or overturned by the violence of the wind.[1] Rudyerd's tower of timber, which at its base was only about twenty-two feet diameter, stood forty-six years before it was burned. It is further remarkable that, while at the building of the Bell Rock, three stones, all dovetailed, wedged, and trenailed, were lifted after they had been permanently set, and at the Carr Rock Beacon twenty-two stones were displaced during its construction, there was no instance of any but loose, unset blocks having been moved during the erection of the Eddystone, although the stones were of very similar weight, and fixed in much the same manner. Smeaton mentions that "after a stone was thus fixed we never in fact had an instance of its having been stirred by any action of the sea whatever."[2]

It is, doubtless, a difficult question to account for the Eddystone tower withstanding so many storms, and at last succumbing to their assaults. It may be that some alteration of the contour of the ledge at the bottom, produced by the erosion of the waves, has changed the direction of their impact on the upper parts of the tower, so as at last to shake the masonry in a way which it did not do at first. But whatever may be the explanation, there can be no doubt of the fact that the masonry was so shaken at last as to lead to its being strengthened, and ultimately to the erection of a new tower.

Additional and very striking corroboration of the influence of the shape of rocks on towers has also been afforded by the experience derived in the construction of the Dhu

[1] An Historical Narrative of the Great and Tremendous Storm which happened Nov. 26, 1703. Lond. 1769. P. 148.

[2] Account of the Eddystone, p. 132.

Heartach Lighthouse. During a summer gale, *fourteen* stones, each of two tons weight, which had been fixed in the tower by joggles and Portland cement, at the level of 37 feet *above high water*, were torn out and swept off into deep water.

It is a remarkable fact (*See* Plate I.) that the level above the sea at which these 14 blocks were removed by a summer gale, is the same as that of the *glass panes* in the lantern of Winstanley's first Eddystone Lighthouse, which, nevertheless, stood successfully through a whole winter's storms. And in his tower, as last constructed, there was at the same level, an open gallery, above which the cupola and lantern were supported on pillars, and this fragile structure stood during four winters. In consequence of the experience of storms, the solid part of Dhu Heartach lighthouse was altered from the original design, and *carried up to the same level above high water as the present lantern in Smeaton's tower.* These facts surely afford sufficient proof of the great influence of the form of rocks on the action of the waves.

The following table from Mr. D. A. Stevenson's account of Dhu Heartach shows the relative levels of the solid at different lighthouse towers :—

Name.	Engineers.	Height of solid portion above high water. Feet	In.	Authority.
Eddystone .	Smeaton	10	3	Smeaton's "Narrative."
Bell Rock .	R. Stevenson	14	0	Stevenson's "Bell Rock."
Wolf . .	J. Walker and J. N. Douglass	16	4	J. N. Douglass, Min. of Pr. of Inst. C. E., vol. xxx.
Bishop . .	J. Walker	23	0	Drawing by J. Walker.
Chickens . .	D. & T. Stevenson	21	0	Reports, D. & T. Stevenson.
Skerryvore .	Alan Stevenson	30	6	Stevenson's Skerryvore.
Dhu Heartach	D. & T. Stevenson	64	4	Reports, D. & T. Stevenson.

The very remarkable cases of wave-action exerted at high levels on the rocks of Whalsey, Unst, and Fastnet, are further corroborative of the same class of phenomena.

The conclusion, then, which is fairly deducible from these facts is, that *the level of the plane of impact of the waves above high water depends upon the relation subsisting between their height and the height and configuration of the rocks above and below high water, as well as on the configuration of the bottom of the sea near the lighthouse.* Thus, while the rock at Dhu Heartach, from its height above high water, forms a great protection against the smaller class of waves, it operates as *a dangerous conductor* to the largest waves, enabling them to exert a powerful horizontal force at a much higher level than they would had the rock been lower. The facts which have been stated do not therefore necessarily prove that the waves are exceptionally high at Dhu Heartach and at the Fastnet, but may be accounted for by the configuration of these rocks. It is no doubt true, as discovered by Mr. D. A. Stevenson,[1] that there is a deep track in the bottom of the sea at Dhu Heartach, extending from the ocean nearly to the rock, and that, therefore, a higher class of waves must reach it than would otherwise have been the case; still the influence of the configuration of the surface is sufficiently obvious both there and at the other places referred to. It is essential that these facts should not be overlooked, for a rock may either shelter a tower from the waves, or, on the other hand, increase their force against it, and cause them to strike higher up than if the rock had been smaller, of a different shape, or at a less elevation above the sea.

13. *Centre of Gravity of the Tower.*—If we except the building erected by Winstanley, of the manner of

[1] Min. of Pro. of the Inst. of Civil Engineers. Vol. xlvi.

destruction of which we do not possess any information, there is no record of a lighthouse of masonry being overturned *en masse* by the force of the waves. And all towers which are fully exposed are necessarily of such dimensions as to resist being overset by the force of the *wind*. Hence the determination of the position of the centre of gravity need hardly be considered in towers of the ordinary construction. But in the one hypothetical case we have referred to, of a cylindric tower being placed on a very small rock and carried high enough to throw sufficient weight upon the lower courses, it *is* possible that the structure may be overset by the combined impact of the waves below and the wind above. The experiments at Dunbar with the marine dynamometer show that on a curved wall the horizontal impact of the waves at 23 feet above high water was only $\frac{1}{84}$th of the vertical upward force with waves of a certain class. If, besides this large reduction of force, we keep in view that on a semicylindric surface it is further very largely reduced, there can hardly be much risk to the finished masonry at 20 or 30 feet above high-water level with waves of ordinary size, when, as in the case we are now considering, the direction of the impinging water is not materially altered by the rocks below. Though I do not wish to speak decidedly where so little is known and where so much yet requires to be known, still I venture to express the opinion that, in so far as relates to the resistance to horizontal force, no jogglings are needed higher than about 30 feet above high water in a cylindrical or nearly cylindrical tower on which the waves strike freely, and the filaments of water are not much altered in direction in the vertical plane by sloping ledges outside of the foundation. But if the slope of these ledges directs the mass of water obliquely, so as to strike far above the bottom of the tower, as at Dhu Heartach or at the Fastnet, additional precautions should certainly be adopted in designing the tower.

14. *Force of Wind on Towers.*—Leonor Fresnel gave the following formula for the force of the wind on towers:—Where S is the stability of the tower, W the weight of a cubic yard of masonry, n the number of cubic yards in tower, b the diameter of base of tower in yards, A the area in square yards, h the height in yards, and f the force of wind per superficial yard—

$$S = \frac{W n \frac{b}{2}}{f \frac{2A}{3} \frac{h}{2}}$$

But in this, as in other formulæ, the strength of the wind is assumed as uniform at all heights. From observations which I lately made, it appears that the force increases vertically in a parabolic curve, whose vertex below the surface of the ground appears to be 72 feet for winds of moderate strength. The formula deduced from these observations, which were made on a pole 50 feet high, is applicable to lighthouses on the land:—If h is the lower point on the tower (not less than 15 feet above the ground), and H that at a higher point, and v and V the respective velocities at those points—

$$V = \sqrt{\frac{H + 72}{h + 72}}\, v$$

In the cases of towers erected in the sea the formula will be probably somewhat different, for the ratio of friction will be less for a yielding surface such as water, than for the solid ground.

15. *Relative Forces of Currents of Wind and Water of Equal Velocity.*—The relative forces of wind and water on the tower are in the proportion of the squares of the velocities multiplied by the specific gravities; so that if V be the velocity of the air, and v the corresponding velocity of the water capable of producing the same force as the air; and if S and s be the specific gravity of sea water and of air—

$$V = v\sqrt{\frac{S}{s}}$$

And if the specific gravity of sea water be taken at 1·028, and air at ·001234 and v be taken as unity, then V = 29·27 times that of moving water required to produce the same force. There are difficulties in dealing with elements so different physically as air and water, but this result may be regarded as at least approximately correct.

16. *Lighthouse Towers of Cement Rubble or Concrete.*—The first rock station at which cement was employed, instead of lime-mortar, was in 1854, at the tower of Unst in Shetland, already referred to. Sir John Coode erected in 1873 the tower of La Corbiere in cement, but it is not exposed to the action of the waves. In 1871 I proposed (*Nature*, September 1871) that monolithic towers of Portland cement rubble might perhaps be employed for towers *exposed to the full force of the waves.* The advantages attending such a plan were stated to be the following:—

"1. The dispensing with all squaring or dressing of materials.

"2. The suitableness for such work of any stone of hard quality, thus rendering it unnecessary to bring large materials from a distance or to open quarries for ashlar.

"3. No powerful machinery for moving or raising heavy materials is required.

"4. A saving in the levelling of the rock for a foundation for the tower.

"5. The ease of landing small fragments of stone on exposed rocks, as compared with the landing of heavy and finely dressed materials."

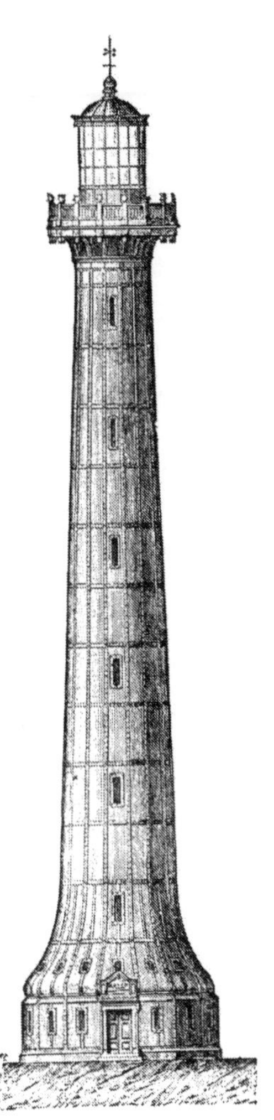

Fig. 15.

How far this system may be available in the erection of lighthouses in the sea is a question which can only be settled by actual trial. A beacon of this construction was erected by Messrs. Stevenson in 1870, in an exposed situation off the Island of Mull, and it has not as yet shown any symptom of decay.

17. *Iron Towers.*—Towers have been constructed both of cast-iron pillars and plates in this country and in America, examples of which are given in Plates VII. and XXI., and in Fig. 15. Some of the remarks which have already been made regarding stone towers may, *mutatis mutandis,* be applied to those of iron.

18. *Mitchell's Screw Pile.*—Where a light has to be placed on a sandbank, the principle of the screw pile, invented by Mr. Mitchell of Belfast, is of great utility, and many examples of its successful application may be found both in this country and in America.

19. *Towers exposed to Floes of Ice.*—The American Board have had to contend with difficulties which are seldom met with in Britain, and never to the same extent. Floes of ice present formidable difficulties to the erection of towers which are exposed to such extreme lateral pressure. Plates No. VII. and VIII. show two American towers, the one constructed of iron and the other of stone, both of which have to resist this peculiar form of assault.

20. *Modes of Uniting the Stones and Courses of Masonry.* —Plates II. and III. show the mode of combining the stones during construction at different lighthouses in this country and in America. The Dhu Heartach mode of connection was suggested by Mr. Alan Brebner, C.E.

CHAPTER II.

LIGHTHOUSE ILLUMINATION.

1. THE best and clearest method of explaining the somewhat intricate arrangements of the various forms of lighthouse apparatus will be to give a continuous historical account, from the earliest period to the present date, of the different optical instruments which have from time to time been invented, so as to show the remarkably gradual development of the different systems of illumination. These successive, and in many respects instructive steps in advancement, extending over a period of more than a century, will, so far as consistent with clearness, be described under each separate head in the order of their publication, or of their employment in lighthouses.

The problem of lighthouse illumination is threefold, and involves to some extent both physical and geometrical optics; but the fundamental principles on which most of the combinations depend, rest really on two or three simple elementary laws of catoptrics and dioptrics. Our attention must be given, 1st, to the source of the light itself, which should produce a flame of constant intensity, and which should, as we shall afterwards see, be of the smallest possible bulk; 2d, Given the source of light, optical apparatus must be designed to collect the greatest possible number of rays coming from the flame, and to direct them to certain parts of the horizon and the sea; and 3d, When lights are multiplied on the same line of coast, it becomes further necessary to introduce distinctions in their character, so that they shall be at once recognisable from each other. One, for example, being constantly in view; another waxing in strength till the full flash is attained, and then waning till complete extinction; a third delivering its greatest power all at once, and then as suddenly eclipsing; a fourth illuminating only a small sector of the horizon; and so forth.

The progress of improvement in the source of illumination from a coal fire to the electric light will be readily conceived; but to understand fully the gradual improvement of the apparatus, it is necessary that some points should first be explained, and these will show that an advance is to be looked for along different lines, and that we must expect to find an ever-increasing success in economising the light by the following means:—1st, By the design of apparatus intercepting more and more of the rays, and directing them with greater accuracy in the required directions; 2d, By reducing the number of optical agents which are employed in producing this result, one new instrument being made capable of doing a greater amount of optical work, so as to effect singly what had formerly required two, thus preventing loss by decreasing the amount of absorption, etc.; 3d, By employing glass rather than metal, because it absorbs fewer rays, and causes less wasteful divergence; and 4th, We shall find improvements in the application of these optical designs to suit the different characters of distinguishing lights, as well as new modifications to meet peculiar wants occasioned by certain geographical features of the line of coast, such as narrow Sounds of varying width, where stronger luminous beams have to be shown in some directions than in others. Lighthouse optics may be said, therefore, to deal with two separate and different problems—viz. *the equal distribution of light either constantly or periodically over the whole horizon,* and *its unequal distribution in different azimuths.* By fully understanding the different objects which are to be attained, the reader will the more readily appreciate the value of the different steps in the march of improvement, from the artless expedient of an open coal fire, or a naked candle, to those optical combinations of maximum power which are geometrically and physically perfect, because they employ the minimum number of agents, consisting of the least absorptive known material, and which condense the whole available sphere of rays diverging from the flame, into one or more beams of parallel rays, or else spread them uniformly, though with different intensities, over those limited sectors of the sea and horizon which alone require to be illuminated. It is no more than a corollary from what has been said, to affirm that *every improvement of lighthouse apparatus may be resolved simply into a method of preventing loss of light;* for, as in mechanics we cannot—as follows from the law of the conservation of energy—bring out more power at one end of a machine than what is supplied at the other, nor even so much, for there must always be a loss due to the friction of the moving surfaces; so in lighthouse apparatus we cannot by any device or combination of appliances bring more light out of the apparatus than that of the unassisted flame, nor so much, because there must always be a loss from absorption and other causes. It therefore follows that the *optimum* form of apparatus is that which does the optical work required of it, by means of the minimum number of glass agents, and produces that amount of divergence only, which is required in each particular case for the guidance of the mariner.

2. *Essential Difference in Power of Fixed and Revolving Lights.*—All light is lost to the sailor which is either allowed to diverge above the horizon, and therefore above the sea, or so far below it as to fall short of the sea. Every apparatus, therefore, must cast its rays as exclusively as possible on the water; because to light up the clouds above the horizon, or the land and the shore below it, can be of no use to navigation. The kind of apparatus to be used will depend on whether it is wished to show a constant fixed light, or one which appears only periodically. If it be the first, then, in order to make it everywhere constantly visible, the natural divergence of the rays all round the horizon must not be interfered with; but those which would pass to the skies should be bent down, and those which would fall on the land and the light-room floor should be bent up. The fixed light apparatus ought therefore to parallelise the rays *in the vertical plane only.* But if the light is to be revolving, it has no longer to illuminate all points of the horizon simultaneously, but only each point of it at successive intervals of time. In addition to the vertical condensation of the fixed light, the periodic flash of a revolving one should obviously be further strengthened by condensing the rays *horizontally, and in all intermediate planes,* thus forming separate *solid* beams or columns of light. Hence, from its more limited requirements, the apparatus for producing the periodic flashes should with the same radiant be enormously more effective than that for producing a fixed characteristic, because the light which, when fixed, is diluted by being spread all round the circle, is in the revolving collected together into one or more dense beams, in each of which all the rays point in the same direction, being horizontal or nearly so.

3. *Metallic Reflection.*—It is a well-known fundamental

law in catoptrics that when rays of light fall on a polished surface they are reflected from it at an angle equal to that of their incidence; but a certain amount of light varying with the angle of incidence, and with the reflective power of the material and its polish, is lost in the act of reflection. This loss is partly due to imperfections in the form or construction of the mirror, which produce dispersion by irregular reflection, and partly to absorption, which arises from the light entering the substance of the metal, and being there transformed into heat. From actual experiments made on silver-plate of the kind generally used in lighthouse apparatus, it was found that at 45° incidence only ·556 of the incident light was reflected.[1]

4. *Refraction and Total Reflection by Glass.*—When rays of light passing through air fall obliquely on the surface of a piece of polished glass, they enter the substance of the glass, but in doing so suffer refraction, being bent out of their original direction, and a certain amount of light is lost, which varies with the angle of incidence. The deviation by refraction where i is the angle of incidence, and r that of refraction, μ the index of refraction of the kind of glass, and δ angle of deviation, is

$$\delta = i - r, \text{ when } \sin r = \frac{\sin i}{\mu}.$$

When the rays, after being refracted at the first surface of a glass prism, reach the second surface at an angle of incidence greater than what is called the *critical* angle, refraction becomes impossible, and they suffer "total" or internal[2] reflection; so that, instead of passing out of the prism into the air, they are sent back again into the substance of the glass. The minimum angle at which this reflection takes place de-

[1] Stevenson's Lighthouse Illumination, 2d Edition: Edin. 1874, p. 12.

[2] Strictly speaking, the light reaches the surrounding air, but does not enter it.

pends on the index of refraction of the glass, its sine being in each case the reciprocal of the index of refraction, which for safety should in calculation be taken for the least refrangible of the red rays of the spectrum. When μ is the index of refraction of the extreme red rays and r the angle of internal incidence,

$$\sin r = \frac{1}{\mu}$$

	Mean Value of μ.	Critical Angle.
Plate Glass . . .	1·510	41° 28′
Crown Glass . . .	1·522	41° 4′
Flint Glass . . .	1·600	38° 41′

Theoretically there should be no light lost by internal reflection (hence called "total"), and Professor Potter has found experimentally that this is the case with very finely polished glass (Phil. Mag., 1832). But in every case light is lost by absorption in passing through the substance of the glass, and by reflection at its surface on entering, and also when the rays again pass out of the glass into the air. The loss in passing through the substance of the glass· increases in a geometric ratio with the length traversed, and is dependent on the degree of homogeneity and colour possessed by the material. If 1 be the whole incident light, q the quantity which escapes absorption in passing through unit of length of a sensibly colourless medium, then the quantity transmitted by n units will be q^n. M. Allard, who has given much attention to this subject, states the loss due to *absorption* in traversing glass, as ·03 per centimetre of length. Professor Potter (Treatise on Optics, 1851) gives the loss by superficial reflection with glass, as $\frac{1}{30}$th for normal incidence. From trials made with a reflecting prism of the kind used in lighthouse apparatus, it appeared that the amount of light transmitted, after passing through it (being 2·6 inches of glass), was ·805 of the whole.[1] From these experiments and those of No. **3**, it follows that by using glass instead of metal there is a saving of about *one-fourth* (·249), thus clearly establishing the superiority of glass over metal as a material for lighthouse purposes. But in addition to this advantage, curves ground in glass certainly admit of much greater accuracy of surface-form than those of the metallic reflectors used in lighthouses. In the one case the result is effected by a gradual process of grinding by means of unerring machinery of rigid and unalterable construction; while in the other it is attained by a comparatively rude tentative manual process, and subject, therefore, to all the imperfections to which such methods of working are obviously liable. The polish too, on which so much depends, is in the glass apparatus given *once for all* by the accurately constructed machinery of the manufacturer; whereas the metallic polish is constantly undergoing deterioration from atmospheric action, and requires to have its brilliancy daily renewed by a succession of different lightkeepers, from the less skilful of whom it may receive ineradicable scratches and permanent injuries. Thus, though the glass while unbroken never loses its correct form, the metallic polish may be deteriorated, and even the curve of the mirror may be altered by an accident.

We shall now proceed to give a historical account of the application of optics to lighthouse illumination:—

I. TO THE ILLUMINATION OF EVERY AZIMUTH, OR OF CERTAIN AZIMUTHS ONLY, BY LIGHT OF EQUAL POWER, ACTING EITHER CONSTANTLY OR PERIODICALLY; and

[1] Lighthouse Illumination, 2d Edition, p. 10. See also Chap. VII.

II. (in the next Chapter) TO THE UNEQUAL ALLOCATION OF THE LIGHT TO DIFFERENT AZIMUTHS, EITHER CONSTANTLY OR PERIODICALLY.

The optical means of producing these different distributions of the light are the CATOPTRIC SYSTEM, acting by metallic reflection only; the DIOPTRIC SYSTEM, where the optical agent is wholly glass; and the CATADIOPTRIC SYSTEM, which is a combination of the two.

Before describing these systems, however, I shall refer, shortly, to some of the earlier modes of illumination, where no optical arrangements were employed.

The first attempt to indicate the position of the sea coast to the sailor at night was

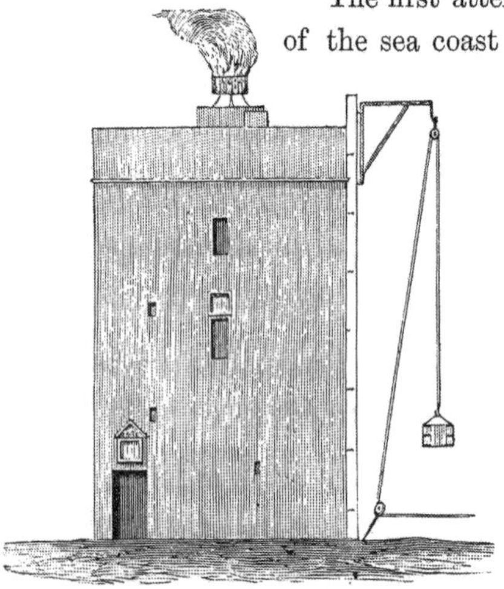

Fig. 16.

10 5 0 10 20 FEET

Fig. 17.

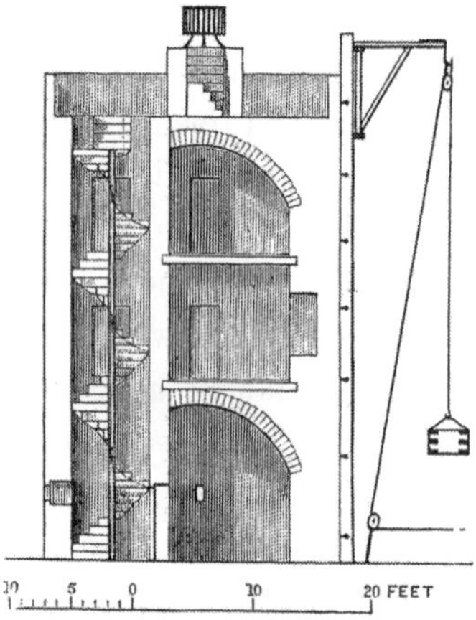

probably by means of a grate of burning wood or coal, placed on the top of a high tower. Figs. 16 and 17 represent one of those beacons, which was erected, in 1635, on the Isle of May, at the entrance of the Firth of Forth. Yet it is somewhat remarkable that the earliest record we possess refers to the use of oil as an illuminant in 1595. In Hakluyt's Voyages, vol. ii. p. 448, it is stated that "at the mouth of the Bosphorus there is a turret of stone upon the mainland, 120 steps high, having a great glass lanthorne in the top, four yards in diameter and three in height, with a great copper pan in the midst to hold oil, with twenty lights in it, and it serveth to give passage into this Strait in the night, to such ships as come from all parts of those seas to Constantinople." The next record we have is of the Eddystone, where in 1696 Winstanley placed tallow candles on a chandelier, also surrounded and protected from the wind and rain by a glazed lantern. It may here be noticed, though in strictness belonging to the dioptric system, that in 1759 a London optician proposed, as Smeaton tells us, to grind the glass of the lantern to a radius of $7\frac{1}{2}$ feet; but the description is too vague to admit of more than conjecture as to the nature of the apparatus which he had in view. The idea was, however, an important one, inasmuch as it contained the germ of the dioptric mode of illumination. Fresnel states that in 1759 lenses were actually used in some English lighthouses, but were in all probability improperly applied, for their use was afterwards abandoned. It was doubtless one of these which is shown in Fig. 18 from a drawing made at Portland lighthouse in 1801. The effect of such an assemblage of fixed lenses would be to throw out narrow beams of light with arcs of darkness between. It is there-

Fig. 18.

fore not surprising that such a partial mode of illumination was discontinued.

Catoptric System of Illuminating every Azimuth with light of equal power, either constantly or periodically.

5. *Parabolic Reflectors.*—It was apparently not till 1763 that optical principles were for the first time correctly applied to lighthouses. Mr. Hutchinson, dockmaster at Liverpool, states in his book on "Practical Seamanship," published in 1777, that lighthouses were erected at the Mersey in 1763; and, at page 180, that they were fitted with reflectors formed of plates of silvered glass, and made, as he says, "as nearly as they can be to the parabolic curve." Figs. 19 and 20

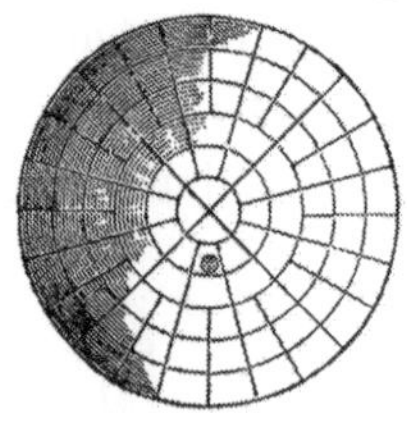

Fig. 19.

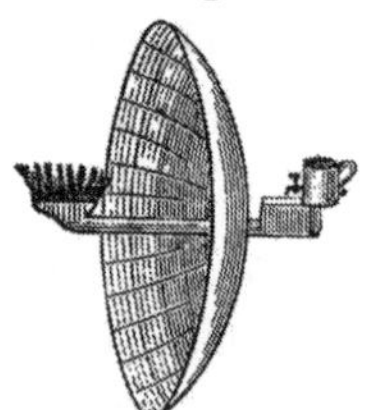

Fig. 20.

show these reflectors as given in Hutchinson's book. This, then, is unquestionably the earliest published notice of the use of parabolic reflectors for lighthouse illumination. They were first introduced into Scotland at Kinnaird Head in 1787, by Mr. Thomas Smith, the first engineer of the Northern Lighthouses, and were also formed of small facets of silvered glass set in plaster of Paris. But in 1804 he substituted silver-plated reflectors at Inchkeith, in the Firth of Forth.

Up to this time the wicks were all of the old flat kind. The ingenious Dr. Robert Hook, so far back as 1677, showed in a monograph, called "Lampas," that an oil flame was in reality a cone of gas, of which the outside only was on fire.

This can be proved by introducing a pipe into the middle of the cone, when the gas will escape and can be burned as a separate light. But it was not till 1782 that Argand carried out what was so nearly suggested in Hook's paper, by making the wicks and burners of a hollow cylindric form, so as to admit a central current of air through the burner; and finally Rumford split up the cone of gas into concentric shells, and ignited them both on the inside and the outside. Peclet, in his Traité d'Éclairage of 1827, states that Argand also used a parabolic mirror, and proposed a revolving light by causing the chandelier to rotate. "It is remarkable," says Mr. J. T. Chance in his excellent Memoir (Min. Inst. Civ. Eng., vol. xxvi.), "how many inventors have contributed their respective parts to the multiple burner—Argand, the double current; Lange, the indispensable contraction of the glass chimney; Carcel, the mechanism for an abundant supply of oil; and Count Rumford, the multiple burner, an idea made feasible by these contrivances, and finally realised by Arago and Augustin Fresnel." Teulère is also said to have proposed the double current burner in connection with the parabolic mirror, and this was applied at Cordouan in France, by Borda, but not till about 1786.

6. *Properties of the Parabolic Reflector.*—These improvements were undoubtedly of great importance, and must have enormously increased the power of a light, as will further appear from the optical properties of the parabola. As any diameter of this conic section is parallel to the axis, and as a normal to the curve at any point bisects the angle between the diameter through that point and the straight line drawn from it to the focus, all rays of light falling on the curve parallel to the axis will be reflected to the focus; and, conversely, all rays diverging from the focus will be

reflected parallel to the axis, so that a luminous point placed in the focus will throw forwards a horizontal column or beam of rays, all of which are parallel to the axis. As, however, the radiant used is not a mathematical point, but an oil light of considerable magnitude, the rays at the outside of the flame are ex-focal, and will after reflection emerge as a cone, whose divergence is dependent on the radius of the flame and the focal distance of the mirror. For most reflectors the useful divergence is about $14\frac{1}{2}°$, and the power is generally taken at 360 times that of the unassisted flame. If Δ be the inclination to the axis of a ray reflected at any point of the mirror, d the distance between the outside of the flame and the focus, e the distance from the point to the focus, $\sin \Delta = \frac{d}{e}$; but as the flame is circular, the total divergence $= 2\,\Delta$.

7. *Useful Divergence.*—The divergence varies therefore, directly as the diameter of the flame, and inversely as the focal distance of the reflector. It follows, then, that the smaller the flame and the larger the apparatus the better, because the incident rays will be better parallelised. Although it is no doubt true that a certain amount of divergence is needed for the sailor, yet the correct practice, which will be afterwards explained, is by means of optical devices, specially designed for each case, to give the exact amount of such divergence, and in the direction only in which it is required. But if we attempt to increase the divergence in azimuth by simply enlarging the flame, we shall *pari passu* increase the divergence in altitude, which will throw more of the light above the horizon, where, as we have said, it is lost, because the sailor cannot see it. It is not therefore going too far to say that for most apparatus all divergence due

to the ex-focal rays of a flame is simply an evil. A bulky radiant is indeed the sole cause of the difficulty that has to be encountered in all attempts to deal accurately with the proper distribution of the light, or to cut it off sharply by means of shades or masks in any one direction in azimuth, for guiding the sailor clear of hidden rocks. Even with the most perfectly constructed dioptric instruments in which an ordinary oil flame is used, if the most intense part of the emergent beam be pointed to the horizon, very nearly one-half of the whole light is lost above the horizon. In France, especially, a different opinion on the subject of divergence has always been, and still is, held by some authorities; but the view now given is optically beyond question, and it may be added that on its truth depends the principal claim which is justly made for the superior qualities of the electric light as an illuminant, and also in a certain degree for the superiority of the dioptric over the catoptric system. It is right to add that as every radiant, however small, is an object of sensible magnitude, the pencil of rays which issues from every kind of apparatus is necessarily divergent, and *its intensity must therefore vary inversely, as the square of the distance from the lighthouse. Optical apparatus cannot therefore abolish the divergence.*

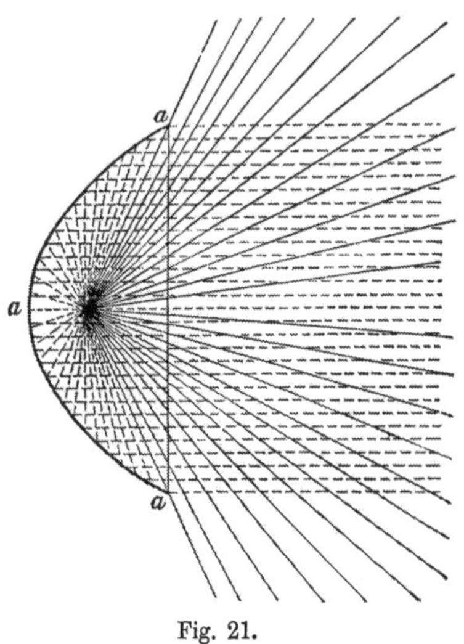

Fig. 21.

8. *Defects of the Paraboloid.*—It will be noticed, from

Fig. 21, that the parabolic mirror *a* is at best but a very imperfect instrument, for *even though the radiant were a mathematical point*, only about $\frac{12}{17}$th of the light would be intercepted by a mirror of the usual form, and the cone of rays shown in Fig. 21, escaping past the lips of the mirror, would be therefore lost.

It is also to be noted that photometric observations show that after reflection the rays are not distributed *uniformly* over the emerging cone, the density of which rapidly decreases towards the edges.

The mode of producing a fixed light on the catoptric system was by arranging a number of reflectors *o* (Fig. 22) around a stationary frame or chandelier *n*. As already

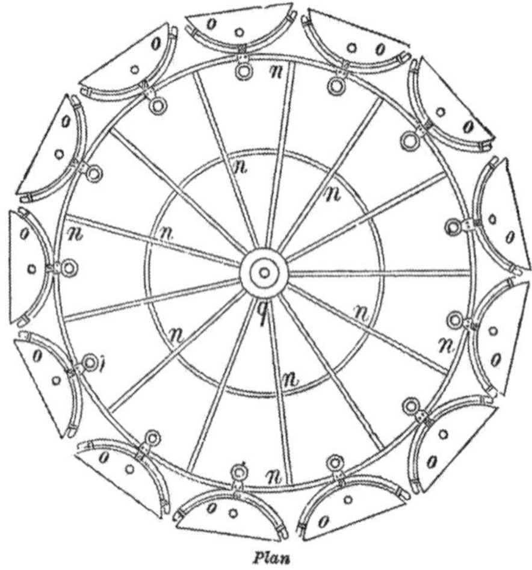

Plan

Fig. 22.

mentioned, an ordinary paraboloid has about 14½ degrees of divergence, so that 25 reflectors were needed to light up continuously (though, as we have seen, not equally) the whole horizon. If, again, the light was to be revolving, a chandelier having 3 or 4 flat faces was employed (*p*, Figs.

23 and 24), on each of which were fixed a certain number of

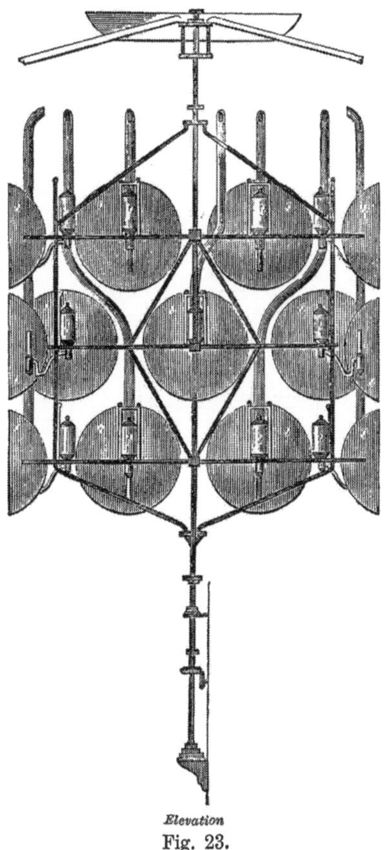

Fig. 23.

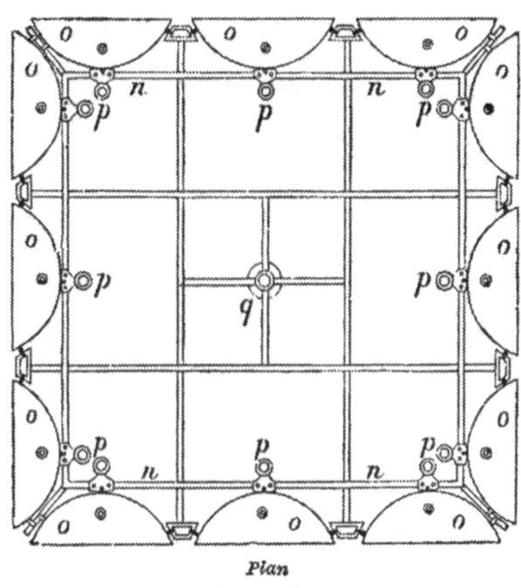

Fig. 24.

separate lamps and reflectors, *o*, having their axes parallel to each other. When the frame was made to revolve, and one of the sides was turned towards the sailor, he would, when at some distance from the shore, receive a flash at once from each of the mirrors which were on that face; but when the face was turned away from him, there would be a dark period until the next face came round. The power of the flashes in this kind of light is obviously proportional to the number of reflectors on each side of the chandelier. If the light was to be flashing, the apparatus was made with eight sides.

9. *Mr. R. Stevenson's focusing arrangement*, 1811.—(Account of Bell Rock, p. 527.) It follows from the laws of reflection that any angular displacement of the position of the flame in relation to the mirror, will produce a corresponding deviation in the emergent rays. It is important, therefore, that after the lamp has been removed to be cleaned or re-wicked, means should be provided for replacing the flame exactly in the focus, so as to be independent of the keeper.

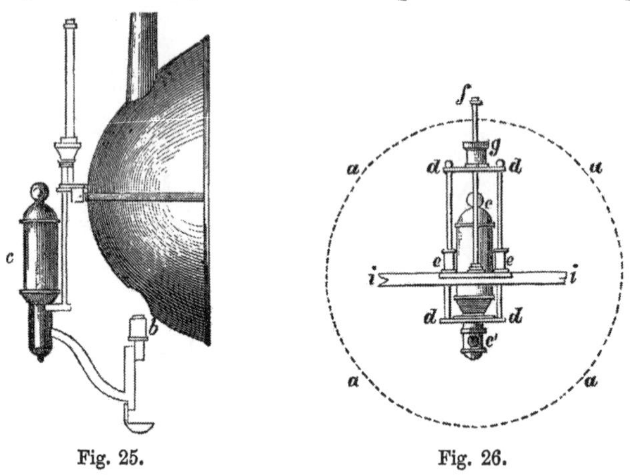

Fig. 25. Fig. 26.

Mr. Stevenson's arrangement for this purpose is shown in Figs. 25 and 26, in which *c* is the fountain for the oil, *b* the burner, *e e* are fixed slides, and *d* and *f* are guide rods. The reflector is shown in elevation in Fig. 25, in which the lamp is represented as lowered down from the reflector. This is effected by the sliding arrangement which ensures its being returned to its true position in the reflector. Sir G. B. Airy, the Astronomer Royal, in his report to the Royal Commission in 1861, says with reference to Girdleness, where this focusing arrangement is employed, "I remarked that by a simple construction, which I have not seen elsewhere, great facility is given for the withdrawal and safe return of the lamps, for adjusting the lamps, and for cleaning the mirrors."

10. *Bordier Marcet's Fanal à double aspect.*—In 1819

Bordier Marcet, in order to reduce the loss of light, invented the instrument shown in Fig. 27. Two parabolas, A C

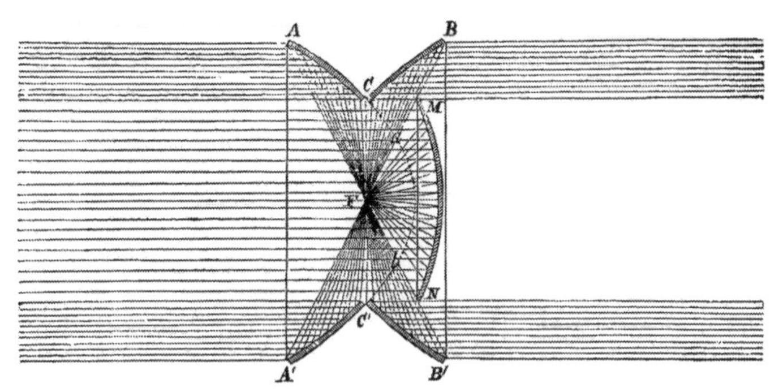

Fig. 27.

C^1 A^1 and B C, C^1 B^1 are placed back to back, so as to point in opposite directions, and a third, M N, is placed within the second, but pointing in the same direction as the first. Though this arrangement possesses some advantages, especially in the smaller divergence, due to the greater focal distance of the mirror M N, as compared with that of the ordinary paraboloid, there is still a large amount of light (the cone A F A′) which wholly escapes interception.

11. *Bordier Marcet's Fanal Sidéral.*—In order strictly to equalise a fixed light over the whole horizon, Marcet also proposed his very ingenious *fanal sidéral* (Fig. 28), which is generated by the revolution of a parabolic profile $p\ p'$ round its parameter as a *vertical* axis, instead

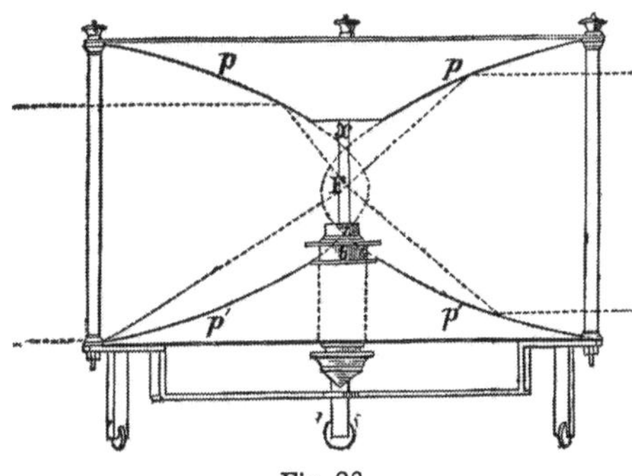

Fig. 28.

of round a *horizontal* axis as in former reflectors. The vertices of the parabolas are cut off, so as to permit of a common focus for the flame. The rays will therefore be reflected by this instrument parallel to the horizontal axis, but in the vertical plane only, while the natural divergence of the light in azimuth will not be interfered with. By this excellent contrivance the light was, for the first time, spread equally round the horizon in one continuous zone. But a large portion of the rays in the vertical plane are still allowed to escape past the lips of the reflector, and this loss takes place all round the circle, even though the radiant were reduced to a mathematical point.

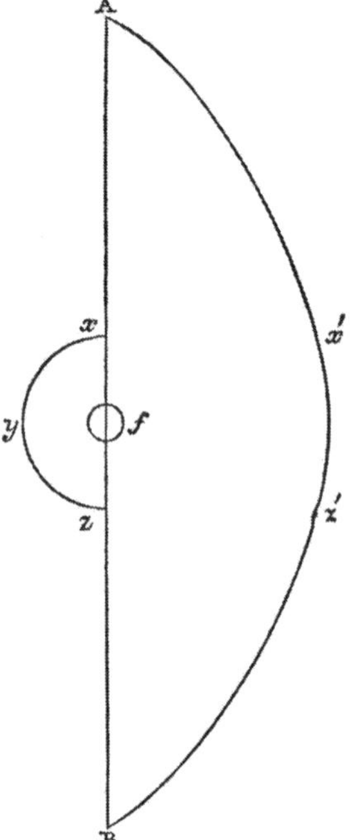

Fig. 29.

12. *Mr. W. F. Barlow's Reflector* (Fig. 29).—In the London Transactions of 1837, Mr. Barlow, F.R.S., in order to intercept more light than by the common parabola, proposed to place in front of that instrument and opposite to the flame *f*, a small spherical mirror *x y z*, so as to catch the cone of rays which would otherwise have escaped without reflection, and send it back through the flame, thence to diverge again and fall upon the parabola behind it. The defects are, 1st, The spherical mirror renders almost useless that part of the parabola *x′ x′* to which it is opposite, and which might therefore as well be removed altogether; but, in addition to this loss, all the rays reflected by the spherical mirror which fall on this useless part of the parabola are also lost. 2d, The necessarily extremely short focal distance of the spherical mirror occasions wasteful divergence in the horizontal, and vertical, and every other plane. 3d, Some of the rays reflected downwards by the spherical mirror are lost by falling on the burner.

Dioptric System.

In the year 1822, which must ever be regarded memorable in the history of lighthouse optics, we come to an entirely new system. As the direction of rays of light can be altered by reflection from a polished metallic mirror, they can also, as we have seen, be refracted into other directions by glass. By reflection they are sent back from the surface on which they impinge, and by refraction they are made to pass through the glass, but in altered directions. The size of the flame produces divergence with refractors, and the amount of this divergence for any point of the lens $= \sin^{-1}\left(\frac{\text{Radius of flame}}{\text{dist. of point from centre of flame}}\right)$.

13. Fresnel's Optical Agents.

1. *Annular Lens.*—So far back as 1748, the celebrated Buffon suggested a new form of lens for *burning* purposes. In order to save the loss of heat by absorption, which must take place with the sun's rays in passing through the thick glass of a large lens whose outer profile is continuously spherical, he proposed to grind out of a solid piece of glass, a lens in steps or concentric zones, so as to reduce to a minimum the thickness, as shown in Fig. 31. This idea was carried into execution by the Abbè Rochon in 1780. Condorcet, the Academician, in his Eloge de Buffon, in 1773 (Paris edition, 1804, p. 35), proposed the capital

improvement of building up Buffon's stepped lens in *separate* rings, so as to correct, to a large extent, the spherical aberration, which is the divergence from the parallel of rays emitted by a lens having a spherical surface. Sir David Brewster in 1811 also described in the Edinburgh Encyclopædia (article "Burning Instruments") the same mode of building lenses and of correcting the aberration. But all these writers designed their lenses for *burning* purposes only, and not with any view to operating on *light*, and still less to lighthouse illumination. In 1822 Augustin Fresnel, celebrated alike as a mathematician and physicist, apparently ignorant both of Buffon's and Condorcet's proposals, described (Mémoire sur un nouveau système d'Éclairage des Phares, 1822), and afterwards constructed a similar lens, but for lighthouse purposes, in which the centres of curvature of the different rings receded from the axis of the instrument according to the distances of those rings from the centre, so as practically to eliminate spherical aberration—the only spherical surface being the central part (*a*, Fig. 30). These lenses were used for revolving lights only.

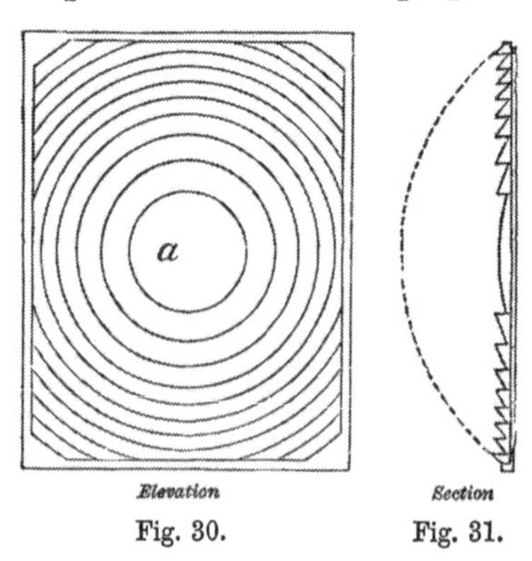

Fig. 30. Fig. 31.

2. *The Cylindric Refractor.*—Fresnel further originated the idea of producing a fixed light by a refracting agent, that should act in the vertical plane only, by bending down the rays that would pass above the horizon, and bending up those that would pass below it, so as, without interfering

with the horizontal divergence, to throw constantly a zone of rays of strictly uniform power over every part of the horizon, thus effecting dioptrically what had been done before by Bordier Marcet's reflector. This cylindric refractor, as it is called, is a zone or hoop of glass (Figs. 32 and 33) generated by the revolution round a *vertical* axis of the middle section of the annular lens already described, which lens, on the other hand, being generated by the revolution of the same lenticular profile round a *horizontal* axis, parallelises the rays in *every* plane.

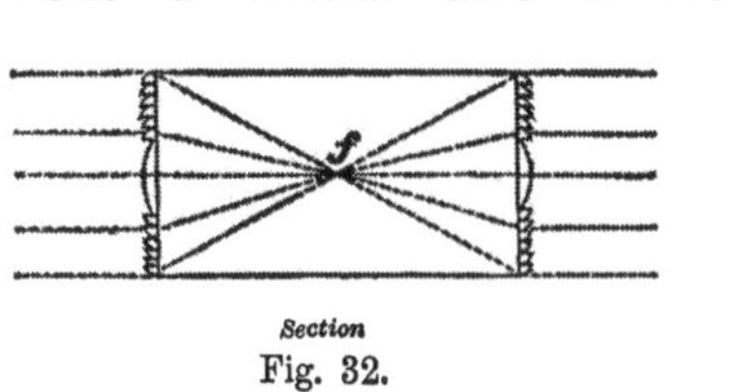

Section
Fig. 32.

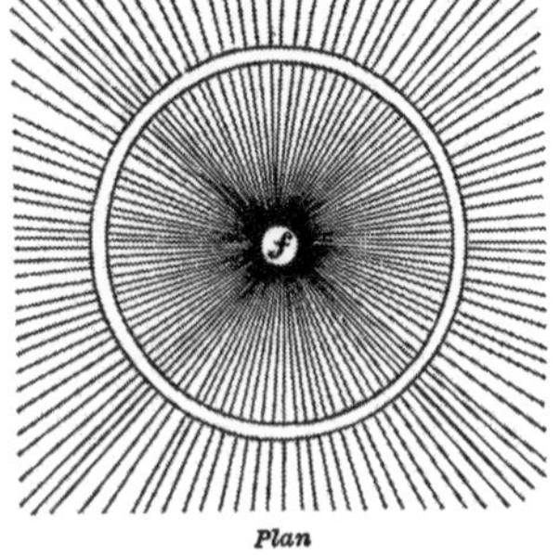

Plan
Fig. 33.

3. *Totally Reflecting Prisms.*—Fresnel next conceived the admirable improvement of employing the principle of "total" or internal reflection by glass prisms. The ray F *i* (Fig. 34), falling on a prismoidal ring, A B C, is refracted and bent in the direction *i* R, and falling on the side A C, at an angle of incidence greater than the critical, is totally reflected in the direction R *e*, and impinging on the side B C at *e*, it undergoes a second refraction, and emerges horizontally. The highest ray F A after refraction by A B and reflection by A C must (in order to avoid superfluous glass) pass along A B, and after a

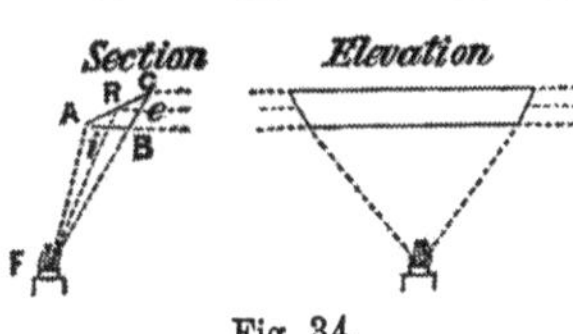

Fig. 34.

second refraction at B emerge horizontally. The lowest ray F B after refraction by A B must, for like reason, pass along B C, and after reflection by A C and a second refraction by B C also emerge horizontally. Every other ray incident on the prism between A and B is, after one reflection and two refractions, emitted horizontally.

4. *The Straight Refracting Prism* is another of Fresnel's optical agents, which refracts the rays that fall on it, but in one plane only. It requires no further explanation, as it is simply a straight prism of the same horizontal cross section as one of the prisms of his cylindric refractor.

14. *Great Central Lamp.*—We will now go on to describe the manner in which Fresnel utilised the four new optical agents which he originated, by first referring to his central burner system. In all lighthouses prior to 1822 the mode of getting up the required power was by employing a sufficient number of separate reflectors, each of which (unless we except Bordier Marcet's mirror) required its own separate lamp. But Fresnel placed in the centre of the apparatus a single lamp, which had four concentric wicks, and was fed with oil by a pump worked by clockwork. Surrounding this burner was a stationary cylindric refractor for a fixed light, and revolving annular lenses for a revolving light.

The mechanical lamp, as it is called, was designed jointly by Fresnel, Mathieu, and Arago, and will be afterwards more fully described.

FRESNEL'S COMBINATIONS OF HIS OPTICAL AGENTS.

15. *Catadioptric Fixed Light.*—This apparatus (Figs. 35, 36) consists of the central lamp, surrounded by a cylindric refractor R, while above and below are zones of silvered mirror A, similar in profile to Bordier

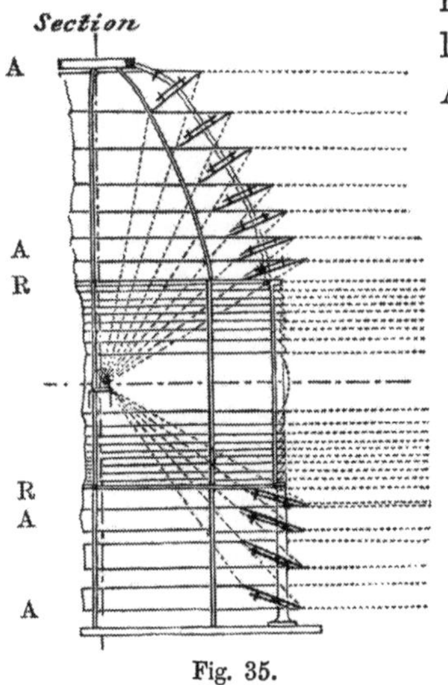

Fig. 35.

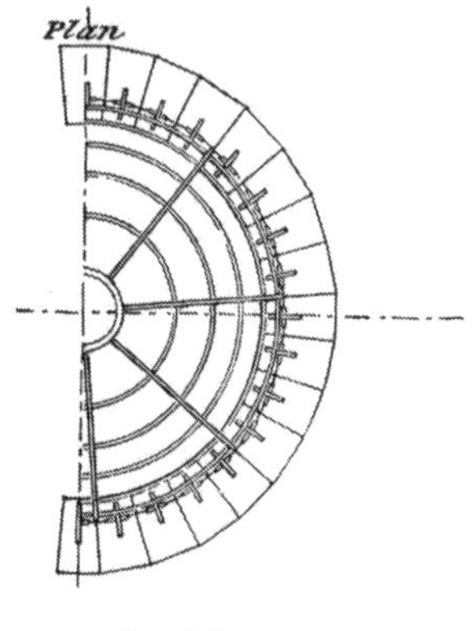

Fig. 36.

Marcet's reflector. By the use of the refractor the whole of the wasteful divergence, which, as we have seen, was the only geometric defect of Marcet's reflector, is entirely prevented, and all the rays are intercepted and spread in a zone of uniform intensity all round the horizon. We have here, then, for the first time in the history of lighthouse optics, a combination which, for the purpose required, is geometrically perfect, though not physically so, because metallic reflection is employed. This defect, as we shall immediately see, Fresnel completely obviated in his next design.

16. *Fresnel's Dioptric Fixed Light. First Application*

of Total Reflection to Fixed Lights.—In Figs. 37 and 38, Fresnel

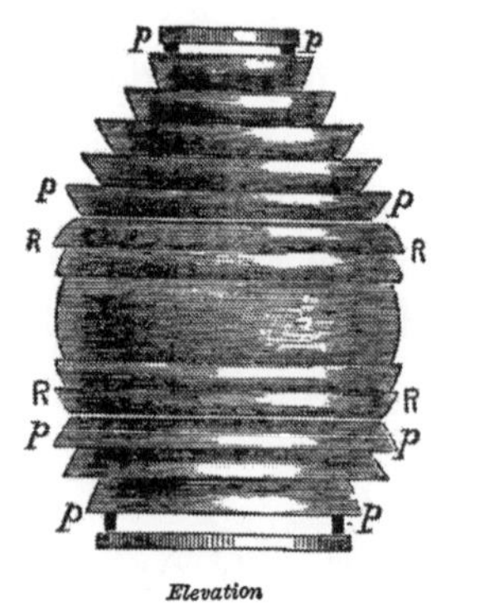

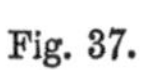

Elevation

Fig. 37.

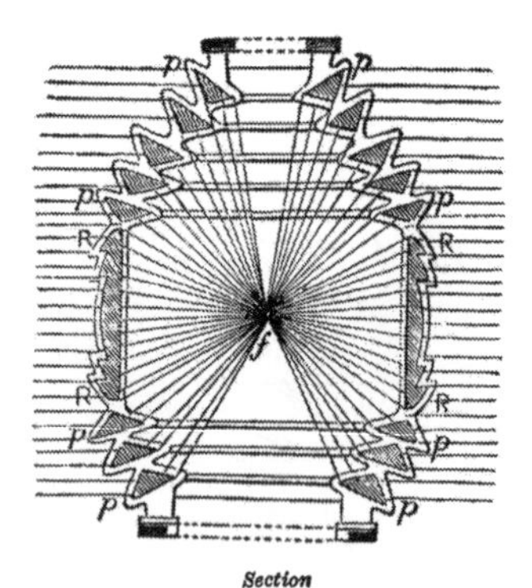

Section

Fig. 38.

substituted his totally reflecting prisms for Marcet's reflectors, so as to spread the light uniformly over the horizon solely by dioptric agents. This was not only the *first application of total reflection to lighthouses,* but was the first optical combination, which, for the purpose required, was both geometrically and physically perfect (excepting of course the inevitable loss due to the divergence of the exfocal rays of the flame), leaving in fact no room for improvement; and, accordingly, this beautiful instrument continues till now in universal use.

In Figs. 37, 38, R R represent the cylindric refractor, *f* the focus, and *p p* the totally reflecting prisms placed at top and bottom.

17. *Fresnel's Revolving Light,* 1822.—In Figs. 39 and 40, which show Fresnel's form of revolving light, the central burner B is surrounded by annular lenses L, and a compound arrangement of inclined trapezoidal lenses l and plane silvered mirrors M. The inclined lenses fit closely to each other, forming a pyramidal dome, and the light, intercepted by them, is sent upwards in beams of a trapezoidal section, until, falling on

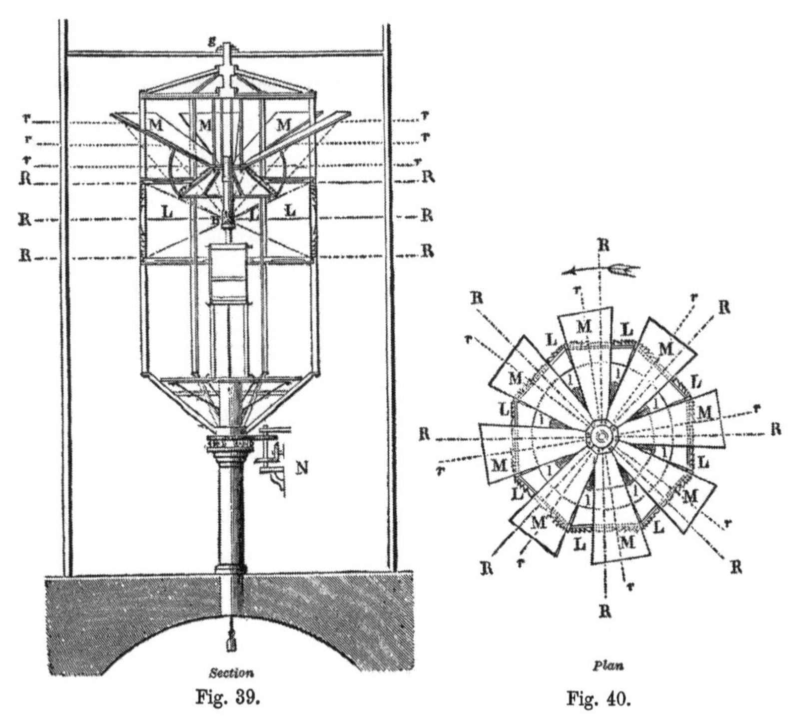

Fig. 39. Fig. 40.

the plane mirrors M, they are bent so far downwards as to emerge with their axes r r parallel to the axes R R of the rectangular lenses below. If, then, all these optical agents are made, by the wheelwork N, to move together round the central lamp B, the sailor will receive the full flash whenever the axes of the emerging beams are pointed towards him, and he will be in darkness when they are turned away from him. Though all the rays are, or as Fresnel pointed out, might be intercepted and sent in the required directions by placing similar inclined lenses and plane mirrors below the central lenses in order to intercept the light which escapes striking on the burner of the lamp, yet the design, unlike that of his fixed light, is very far from being perfect; not only because metallic reflection is employed, but because two agents are needed for all but the central portion of the rays, thus causing a large loss, to which Fresnel himself was fully alive, as appears from this passage in his Memoire of 1822 (p. 17):—"Mais comme on est obligé d'employer des glaces etamées pour raméner dans une direction horizontale les faisceaux lumineux qui sortent de ces lentilles, une grande partie de la lumiere incidente est absorbée par les miroirs, malgré leur inclinaison prononcée, qui est de 25°; *et jéstime que la lumiere incidente doit être réduite à moitié par son passage au travers des lentilles et sa réflexion sur ces glaces étamées.*" The first light on this principle was that of Cordouan in 1822.

18. *Fresnel's Fixed Light, varied by Flashes.*—This distinction (Figs. 41 and 42), Fresnel produced by placing his straight refracting prisms r' (**13**, 4) on a revolving frame outside of his fixed light apparatus $p\ x$. Now, as that apparatus parallelises the rays in the vertical plane only,

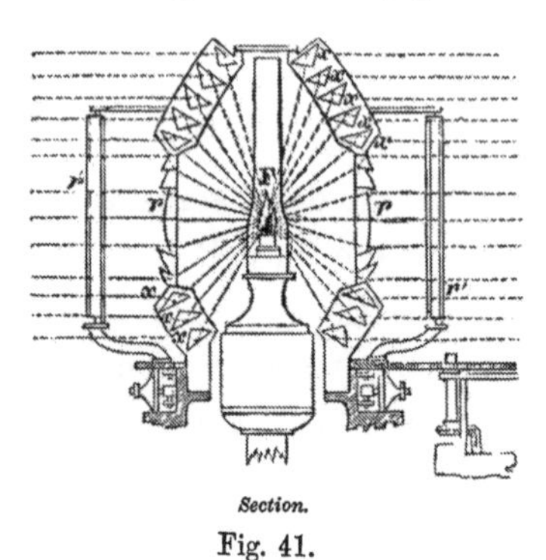

Section.

Fig. 41.

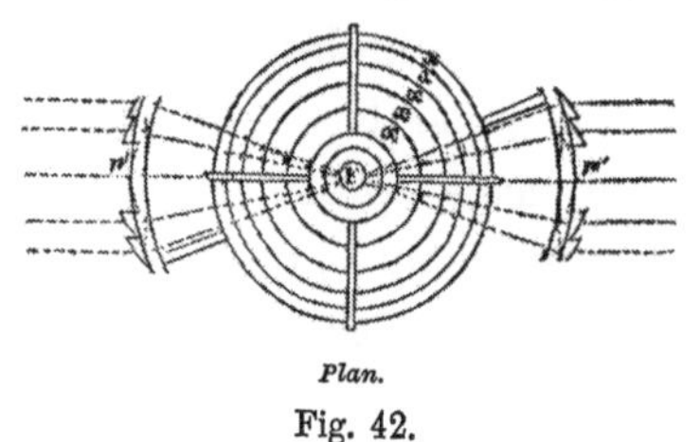

Plan.

Fig. 42.

and the straight refractor in the horizontal plane only, it is obvious that whenever the revolving straight prisms come in line with a distant observer, the light must increase enormously in volume, giving, instead of a narrow strip of light, a broad flash like an ordinary lens. The defect in this apparatus, although glass only is used, is the use of two agents for producing the effect.

19. *Sir David Brewster's Catadioptric Arrangement.*— In 1823 (Edinburgh Philosophical Journal) Sir David proposed to apply his burning arrangement of 1811 to lighthouse illumination. The only difference between the design by Fresnel which was the first that was made for lighthouses, and this proposal of Brewster was, that the latter showed an arrangement for collecting the rays into a single beam, and employed a spherical mirror *m n*, for dealing with the back rays.[1] In Fig. 43, *a* is the main lens; *c, b, d, e,* the inclined lenses; *m n*, the spherical mirror; and *t u*, *p q*, *r s*, and *v w*, the plane silvered mirrors; and *f* the flame. The defects are the same as have been already stated in relation to Fresnel's plan.

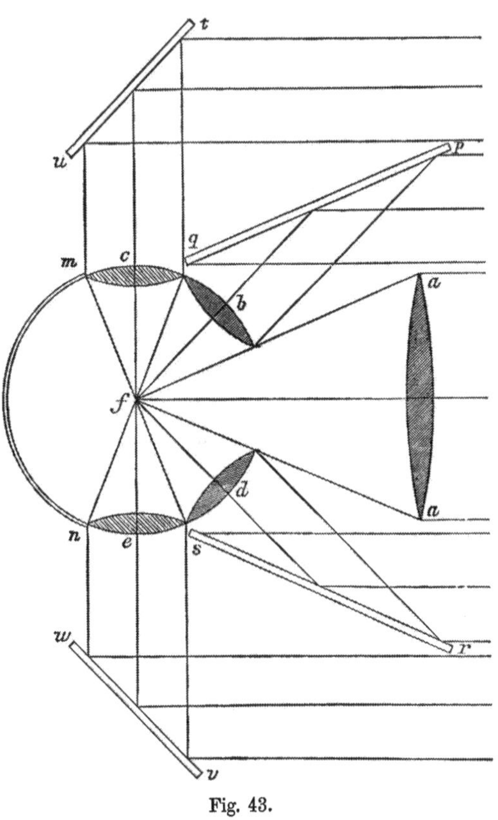

Fig. 43.

[1] The spherical mirror had previously (1790) been used for lighthouses by Thomas Rogers.

20. *Introduction of Dioptric System.*—For a long period after the lamented death of Augustin Fresnel in 1827, his brother, M. Leonor Fresnel, who succeeded him as Lighthouse Engineer, was largely instrumental in introducing these most valuable improvements into the lighthouses of the world. The Dutch are believed to have the honour, after France, of having been the first to introduce the new system. Next in order come the Scotch Board, who, in 1824, sent Mr. Robert Stevenson to Paris, and he recommended its employment at Buchanness Lighthouse in 1825. This proposal, however, was not carried out, as a different character of light was adopted for this headland. In 1834 his successor, Mr. Alan Stevenson, visited Paris, where, through the liberality of L. Fresnel, he received the fullest information as to his brother's system, which he embodied in an elaborate report printed at Edinburgh in the same year; and in 1835 the revolving light of Inchkeith, and in 1836 the fixed light of Isle of May, were made dioptric instead of catoptric. Mr. Stevenson also published Fresnel's formulæ for calculating the different dioptric instruments, and he computed and published the elements of a complete cupola of totally reflecting prisms. The Trinity House of London followed next, and employed Mr. Stevenson to superintend the construction of a first-order revolving light, which was afterwards erected at the Start Point, in Devonshire.

The Americans, considering their well-known enlightenment and deep interest in naval matters, were somewhat slow in the matter of coast illumination. It seems to have been more than twenty years after the invention of the dioptric system in France that the subject was seriously taken up. In 1845 they applied to M. Leonor Fresnel for full information, and in an elaborate report (December 31st,

1845), which he furnished to the American authorities, he concluded:—"1st, That the lights fitted with dioptric apparatus present a variety in their power and in their effects, and may be made to produce an intensity of lustre which renders them of an interest, in a nautical point of view, incontestably superior to those fitted with catoptric apparatus. 2d, That, if we take into account the first cost of construction and the expense of their maintenance, we shall find, in respect to the effect produced, the new system is still from *once and a half to twice as advantageous as the old*."[1] The Report also contained the results of photometric observations of the powers of the different instruments then known, which clearly proved the superiority of the dioptric system. It was not, however, until 1852, that a committee of the American Board resolved —"That the Fresnel or lens system, modified in special cases by the Holophotal apparatus of Mr. Thomas Stevenson, be adopted as the illuminating apparatus of the United States, to embrace all new lights now or hereafter authorised, and all lights requiring to be renovated either by reason of deficient power or defective apparatus."

21. *Mr. Alan Stevenson's Improvements.*—When establishing the new system in Scotland, Mr. Stevenson made the following improvements:—

1. *Refractor of a truly Cylindric Form.*—Owing to difficulties in construction Fresnel had to adopt a polygonal instead of a cylindric form for his refractor. But for the Isle of May light, in 1836, Mr. Stevenson succeeded in getting Messrs. Cookson of Newcastle to construct a first-order refractor of a truly cylindric form.

[1] Report of the American Lighthouse Board, 1852, p. 602.

2. *Helical Glass Joints for Fixed Lights* (Figs. 44, 45).—

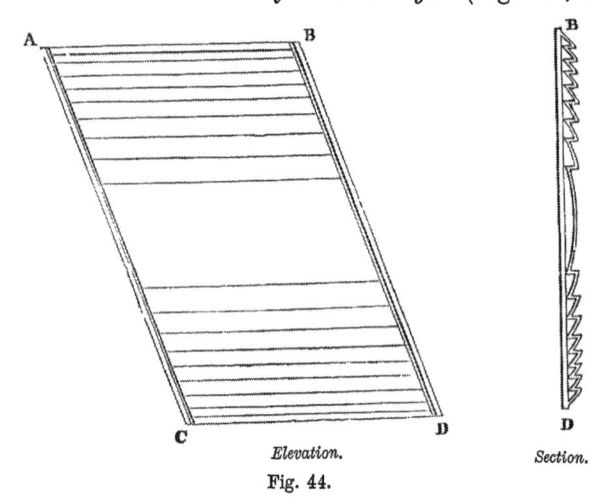

Elevation. *Section.*

Fig. 44.

Mr. Stevenson also introduced the valuable improvement of constructing the refractor in rhomboidal, instead of rectangular pieces. In this way the joints of the glass, A B C D (Fig. 44), instead of being upright, were helical, and the loss of light at the joints was better distributed, and prevented obscuration in any one direction. The amount of light intercepted in any azimuth is inversely proportional to the sine of the angle of inclination of the joint.

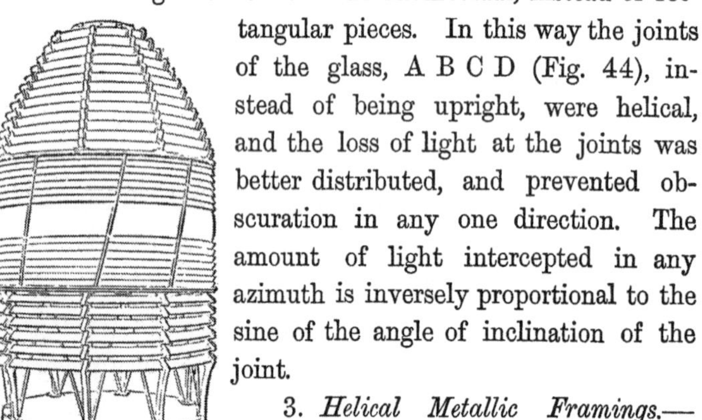
Fig. 45.

3. *Helical Metallic Framings.*—The internal metallic framings for supporting the upper cupola of prisms were also, in like manner, and for the same reason, made by him of a helical form, as seen at the bottom of Fig. 45.

4. *Diagonal Lantern.*—The astragals of the lantern were, for the same purpose, made diagonal, and were constructed of bronze instead of iron, in order to reduce their sectional area (as shown in Plate XI.)[1] A small harbour light, with inclined astragals, was made by Mr. E. Sang about 1836.

22. *Improved Revolving Light of Skerryvore.*—In 1835 Mr. Stevenson, in a Report to the Northern Lighthouse Board, proposed to add fixed reflecting prisms (*p*, Fig. 46) below the lenses of Fresnel's revolving light; and he communicated this proposal to M. L. Fresnel, who approved of his suggestion, and assisted in carrying out the design in 1843. This combination added, however, but little to the power of the flash, and produced both a periodically flashing and a constant fixed light, but it must be remembered that the fixed form was the only kind of reflecting prism then known. The trapezoidal lenses and plane mirrors were also, for the same reason, still used. The prisms for Skerryvore were the first that were made of the large size (first order), and were constructed by M. Soleil at Paris, under the immediate superintendence of M. L. Fresnel.

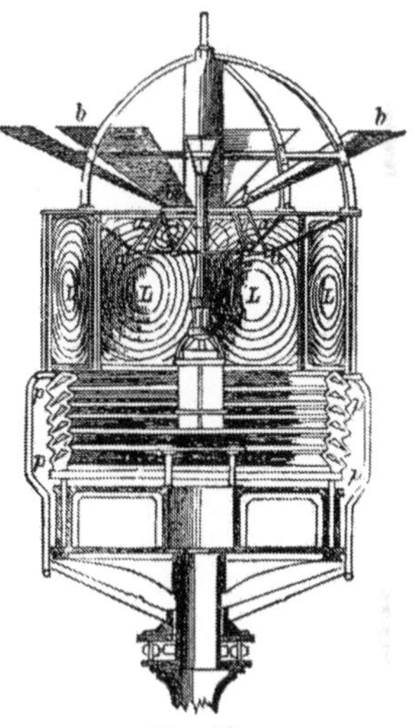

Fig. 46.

[1] Mr. A. Stevenson also extended the helical principle to the astragals of the lantern in a design for Start Point, in Orkney, in 1846, but it was not adopted. Mr. Douglass, independently, made a similar proposal in 1864, and was the first to publish it and introduce it into lighthouses. Mr. Douglass's lantern will be described subsequently, and is shown on Plate XII.

23. *Mr. A. Gordon's Catadioptric Reflector* (Civ. Eng. Jour., 1847).—The next improvement was for increasing the power of the parabolic reflector dioptrically (Figs. 47,

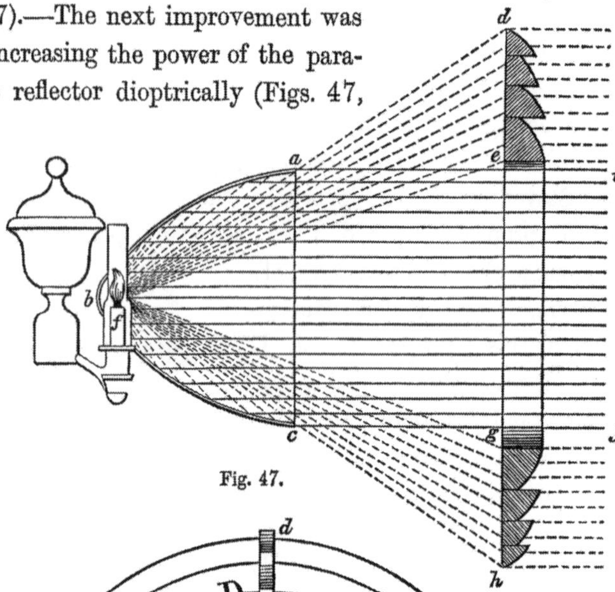

Fig. 47.

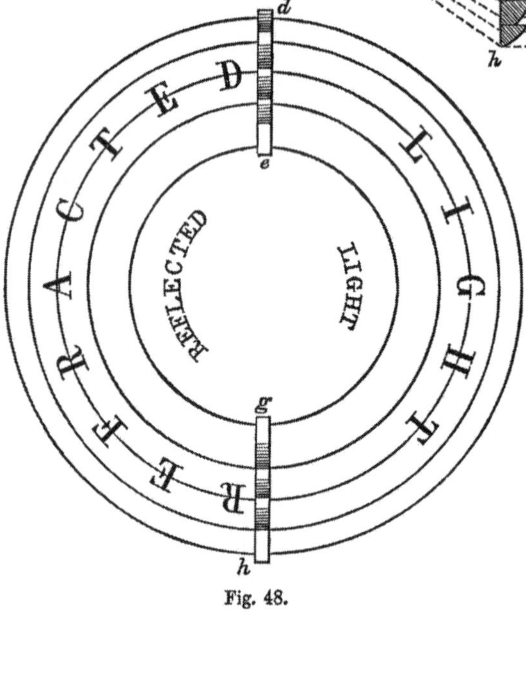

Fig. 48.

48). Mr. Gordon placed in front of a paraboloid, *a b c*, four of the outer rings of Fresnel's annular lens, *d e g h*, for the purpose of intercepting some of the rays which escape the action of the reflector, while the beam from the paraboloid itself passed through the central void, *e g*, between the lenticular rings. Though so far a step in the right direction, the defects are very obvious,—1st, The escape of the cone of rays *e f g* through the central void. 2d, The extremely short focal distance of the reflector, which was found on trial to produce a cone having a divergence of no less than 60°; and 3d, The loss by metallic reflection from the paraboloid.

24. *Leonor Fresnel's improved Revolving Light.*—M. L. Fresnel published an improvement on his brother's light, in his "Phares et Fanaux des Côtes de France, Paris, 1842," by enlarging and altering the position of the trapezoidal inclined lenses (Fig. 49), and thus increasing their focal distance. A very considerable improvement was thus effected, but he still retained metallic reflection and double agents for the upper rays.

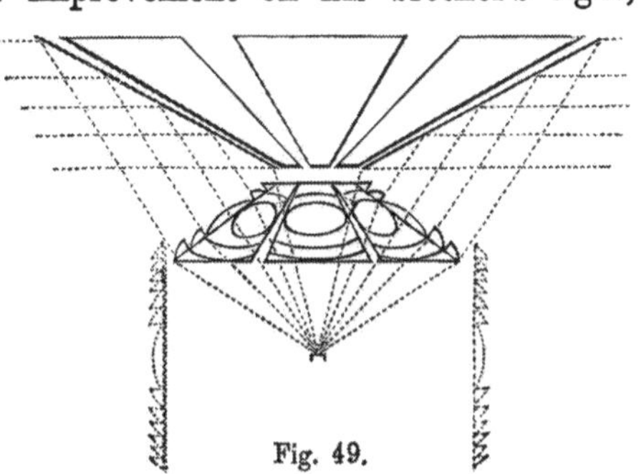
Fig. 49.

25. *Holophotal System*—These improvements come next in order of date, but for the sake of clearness and unity we shall first describe the remaining French improvements on Fresnel's revolving light.

26. *M. Lepaute's form of Revolving Light.*—M. Lepaute, the collaborateur of A. Fresnel, gave a design in 1851, in which, in order to avoid the use of double agents, he increased the height of the lens, and reduced proportionally the angle subtended by the fixed light prisms above and below. In this way he extended the powerful part of the light probably farther than was consistent with favourable angles of incidence of rays falling near the top and bottom of such elongated lenses. The apparatus could therefore parallelise the rays in the vertical plane only by its upper and lower prisms. Of course, if he had been acquainted with the holophotal prisms subsequently to be described, he could have parallelised the light in every plane from top to bottom of the apparatus. In M. Lepaute's letter to the U.S. Lighthouse Board, of 28th July 1851, he states that his design "received the approbation of the Commissioners of Lights in France;" and he adds that "The French Administration is about to order from the undersigned an apparatus of the first order of this description of flashes for every minute, to renew the apparatus of the light of Ailly, near Dieppe."

27. *Letourneau's improved Fixed Light, varied by Flashes.*—Fig. 50 shows in section another purely dioptric apparatus, which was exhibited by Letourneau and Wilkins at the great London Exhibition (Illustrated Exhibitor, 1851). The central drum of this apparatus consisted of alternate panels of the cylindric refractor R R, and the annular lens L L, which last is surmounted by straight refracting prisms *b b*, all of which agents revolve together; while above and below the lenses and refractors

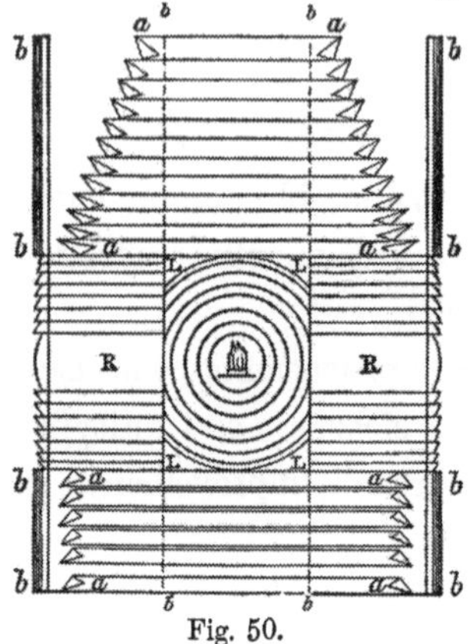

Fig. 50.

are portions of fixed light prisms *a a*, which are stationary. So long as the cylindric refractor is opposite to the observer the light appears fixed, but when the combination of lenses and straight prisms comes round a solid beam of light is produced singly by the central lenses, and doubly, as in Fresnel's original arrangement (No. **18**), by the fixed horizontal prisms and revolving straight prisms. The advantage here gained is no doubt very considerable, for the whole of the central light is parallelised by single agents, but there is still the defect of employing double agents for the upper and lower divisions of the apparatus which produce the flashes.

28. *Improved Fixed Light, varied by Flashes, as constructed by M. Tabouret.*—This apparatus (Fig. 51) was exhi-

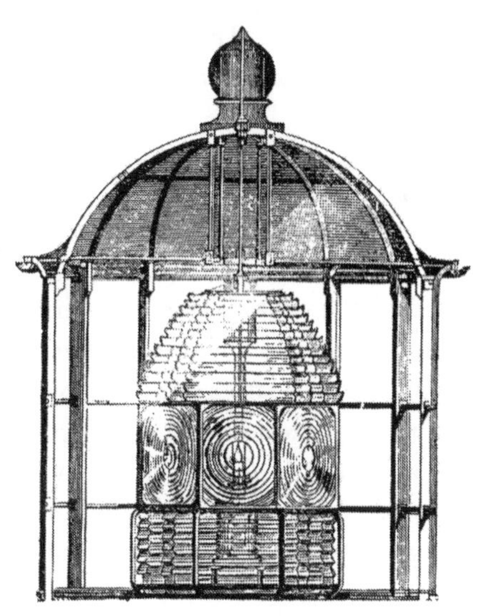

Fig. 51.

bited at the London Exhibition of 1851 by Messrs. Chance, and was made under the superintendence of M. Tabouret, who constructed all the apparatus for Augustin Fresnel till his death in 1827. The object of the design was to equalise more nearly than was done at Skerryvore (**22**), the fixed and revolving beams, and for this purpose the fixed prisms were introduced above, as well as below the middle part of the apparatus, as in Lepaute's design, so as to supersede Fresnel's compound arrangement of inclined lenses and mirrors. Thus, although double agency was avoided, the power was very largely reduced, for only the central part of the luminous sphere was parallelised horizontally and vertically.

29. *Reynaud's Improvement*, 1851.—This was a design in which revolving straight prisms were placed in front of all the fixed prisms, for the purpose of increasing the power of the lenses by suppressing altogether the fixed light which was employed at Skerryvore. The effect would no doubt be a great addition of power, but it was obtained by means of double agents.

Holophotal System, 1849.

30. The object of the different improvements on Fresnel's revolving light which have been described, was to obtain the best utilisation of all the rays. This was perfectly attained without unnecessary agents by the system now to be explained. It will be seen from what follows that by its agency the *revolving* apparatus is rendered optically as perfect as the fixed, and further, that, from the wideness of the application of the principle, its introduction has led to many and various changes in other forms both of catadioptric and dioptric apparatus.

1. *The Catadioptric Holophote.*—The holophotal arrangements which I proposed in 1849-50 (Trans. Scot. Royal Soc. Arts, vol. iv.) show the modes of solving the problem of *con-*

densing the whole sphere of diverging rays into a single beam of parallel rays, without any unnecessary reflections or refractions. Part of the anterior hemisphere of rays (Figs. 52, 53) is intercepted and at once parallelised by the lens *L*,

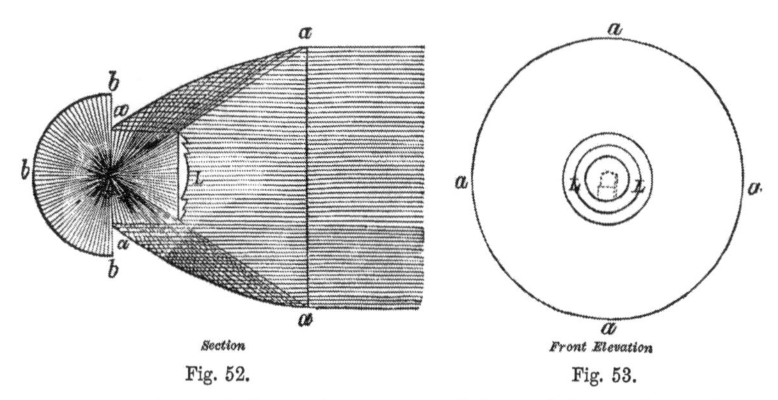

Fig. 52. Fig. 53.

whose principal focus (*i.e.* for parallel rays) is in the centre of the flame, while the remainder is intercepted and made parallel by the paraboloid *a*, and thus the double agents in Fresnel's and Brewster's designs (**17, 19**) are dispensed with. The rays of the posterior hemisphere are reflected by the spherical mirror *b*, back again through the focus, whence passing onwards one portion of them falls on the lens and the rest on the paraboloid, so as finally to emerge in union with, and parallel to the front rays. This was the first instrument which intercepted and parallelised all the rays proceeding from a focal point by the minimum number of agents. It is therefore geometrically perfect, and hence called a *Holophote.*[1] But it is not physically so, for it employs metallic reflection; and with an ordinary oil flame and burner many of the rays reflected by the spherical mirror would, as in Barlow's design, fall upon the burner

[1] I adopted this name from the Greek words ὅλος and φῶς.

and be lost. This instrument was employed at the North Harbour of Peterhead in 1849, which was *the first light in which all the rays were geometrically combined in a single beam, without unnecessary agents.* Figs. 54 and 55 show another arrangement on the same principle, in which the

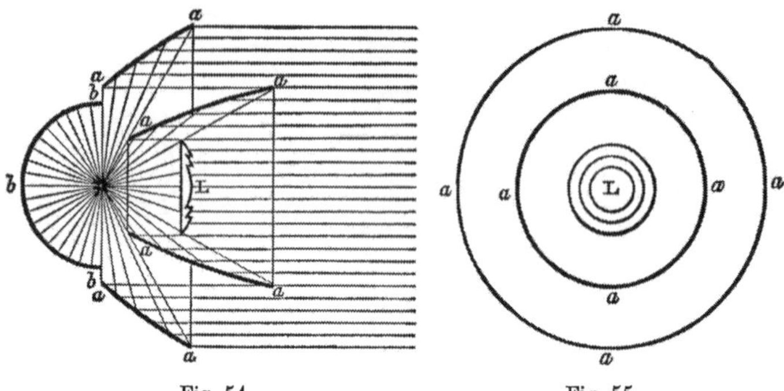

Fig. 54. Fig. 55.

apparatus is made more compact; and Figs. 56 and 57 show a method of utilising the back part of the light, so as to avoid the loss of light due to reflected rays falling on the burner. The prisms (p) will be described in No. 3.

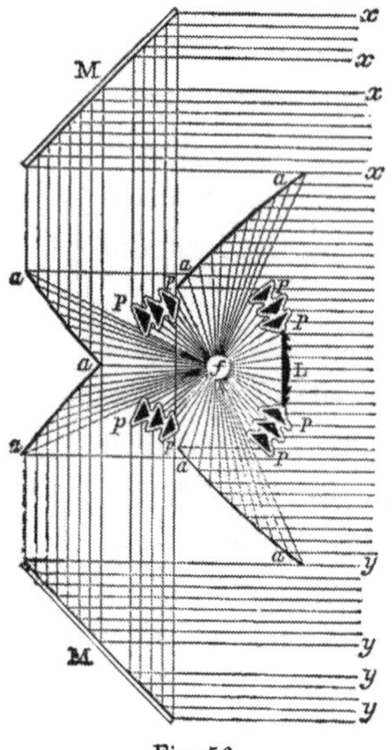

Fig. 57.

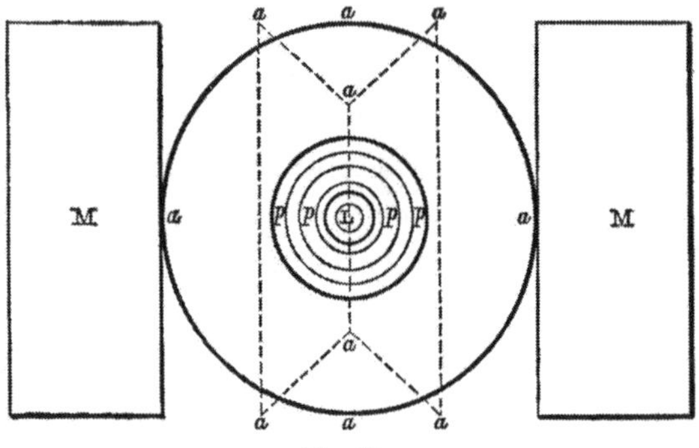

Fig. 56.

2. *Holophotal Catadioptric Apparatus revolving round a Single Central Burner.*—The application of the holophotal principle to catadioptric revolving lights with a central burner is given in Fig. 58. In place of Fresnel's compound arrangement of trapezoidal lenses and plane mirrors (**17**), the mirrors *R R*, instead of being plane, are generated by a parabolic profile passing round a horizontal axis, and all rays from the focus are thus at once parallelised by a single agent. The design is therefore geometrically perfect, like that last described; but metallic agency is still employed. It was first adopted at Little Ross, one of the Northern Lighthouse stations.

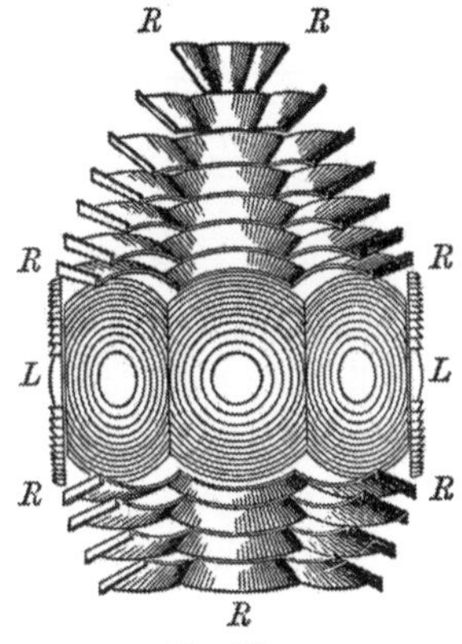

Fig. 58.

OPTICAL AGENTS.

3. *Holophotal Prisms.*—The holophotal prisms (Trans. Roy. Scot. Soc. of Arts, 1850), though similar in section to those in Fresnel's fixed light, differ in the mode of their generation and in their optical effect, for the sections are made to revolve round a *horizontal* instead of a vertical axis, and the incident rays are therefore made parallel in every plane instead of in the vertical plane only. True lenticular action is thus extended by *total reflection* from 45°, which is about the limit of Fresnel's refracting lens, to nearly 130°.[1]

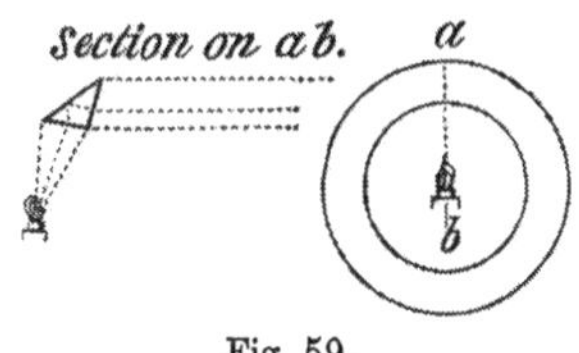

Fig. 59.

[1] In 1852 it was for the first time stated that, so far back as 1826, similar prisms had been made for A. Fresnel by M. Tabouret, and tried, though

4. *Double Reflecting Prisms.*—In 1850 (Trans. Roy. Scot. Soc. Arts) I proposed prisms for giving two internal reflections instead of one. The surface *b c* (Fig. 60) is concave, the centre of curvature being in the centre of the flame *f*. The other surfaces, *a b* and *a c*, are portions of parabolas, whose common focus is *f*. A ray diverging from *f* will fall normal to the surface *b c*, at *e*, and therefore will pass on without any deviation, or loss by superficial reflection at its incidence on *a b*, where it is totally reflected in a direction *r r′* at right angles to the axis of each zone. At *r′* it is again reflected, and finally emerges in a radial

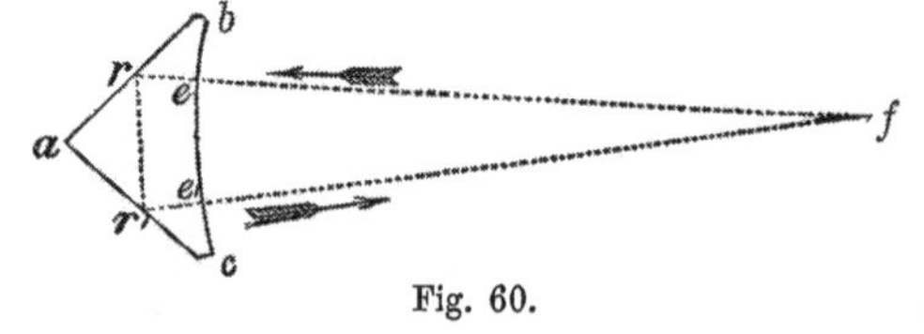

Fig. 60.

unsuccessfully, on lamp posts for lighting the quays of a canal in Paris, and, as is admitted by M. Reynaud, they were not designed in such a way as to be applicable for lighthouses.* *It is also admitted that no drawing or description of them was ever published, and that they were never used in any lighthouse.* But the following facts incontestably prove that even their existence was unknown, or entirely forgotten, for *quarter of a century* after 1826, both by Leonor Fresnel (Augustin's brother and successor) and by Tabouret, who is said to have made the canal apparatus :—1st, Alan Stevenson in 1834, before reporting on the dioptric system, was for nearly a month almost daily in L. Fresnel's office and in Tabouret's workshop, but never heard of them then, nor at any of his frequent visits to Paris afterwards ;† 2d, In 1842 L. Fresnel published his own improvements on his brother's plan (24), and in 1843 superintended the apparatus of Skerryvore (22) ; 3d, In 1850 I went to Paris for the express purpose of showing L. Fresnel a model of my dioptric holophotal arrangement, and he did not at either of two interviews question its novelty ; 4th, All apparatus, even as small as 7 inches radius, where there could be no difficulty of construction, continued to be made at Paris with unnecessary agents ; 5th, In 1851 Tabouret himself constructed and exhibited at the London Exposition his own apparatus (28) ; 6th, In 1851 Lepaute wrote that the French Lighthouse Commissioners approved of and "were about to order"

* Annales des Ponts et Chaussées, 1855.

† Report to Northern Lights, Edin. 1834. Reply to Letter from L. Fresnel to Arago by T. Stevenson, Edin. 1855.

direction at the point *e*. A ray on the other side of the axis will be simultaneously reflected in the opposite direction, and also sent back to the flame. The mode of employing such prisms will be given in next section.

Optical Combinations of Dioptric Holophotal Agents.

5. *Dioptric Holophote with Dioptric Spherical Mirror.*—If holophotal rings (*p p*, Fig. 61), with a central refracting lens, *L*, together subtending 180°, are placed before a flame, the front half of the diverging sphere of rays will be at once condensed by refraction and total reflection into a beam of parallel rays, and the back half may be returned through the flame by a metallic spherical mirror *b*. If, instead of the metallic spherical mirror, a dome of glass (Fig. 62), formed of zones generated by the revolution of the cross section of the double reflecting prisms (Fig. 60) round a horizontal axis, be placed behind the flame, then the back hemisphere of diverging rays falling on it will all be returned through the flame, so as to diverge along with the front rays, for this dome is a perfect spherical mirror not only

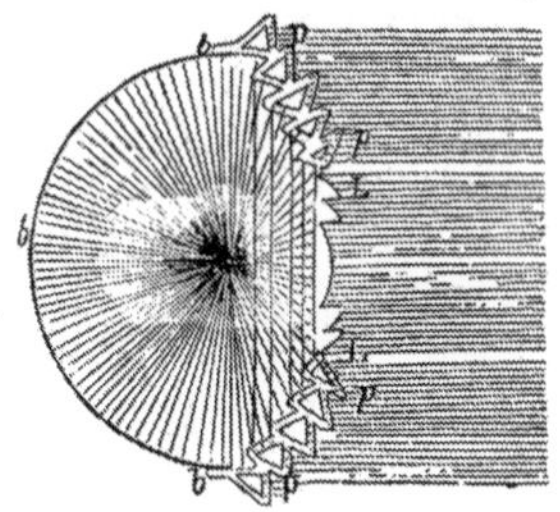

Fig. 61.

his form of apparatus for Cape Ailly, and in the same year they invited tenders (since published*) for that lighthouse on his plan, which were recalled after they had seen the Scotch holophotal apparatus then being made in the workshop of M. Letourneau in Paris;† 7th, In the same year M. Degrande, one of the Government lighthouse engineers, claimed for himself the invention of the holophotal prisms, not having then, as he afterwards told me, heard anything either of the canal prisms, or of my prior publication.‡

* De l'application du principe de la Réflexion Totale aux Phares Tournants. V. Masson. Paris, 1856.

† *Ibid.*

‡ Reply to Fresnel's Letter to Arago, 1855.

for that portion of the faint light that is superficially reflected by its inner surface, but also by total reflection for the much greater portion which enters the substance of the glass; so that all the light is reflected back again through the flame, and finally falls on the parallelising agents in front of the flame. Thus *the whole light is parallelised entirely by glass, metallic reflection being wholly dispensed with in every part of the apparatus, and refraction and total reflection substituted.* It has further to be noticed that this mirror does not concentrate the heat on the burner, as was pointed out by Professor Swan. A thermometer placed behind the mirror stood, after 24 hours, 5° higher than the surrounding air, showing that the heat rays are largely transmitted through the glass, and thus the oil in the burner is kept comparatively cool.

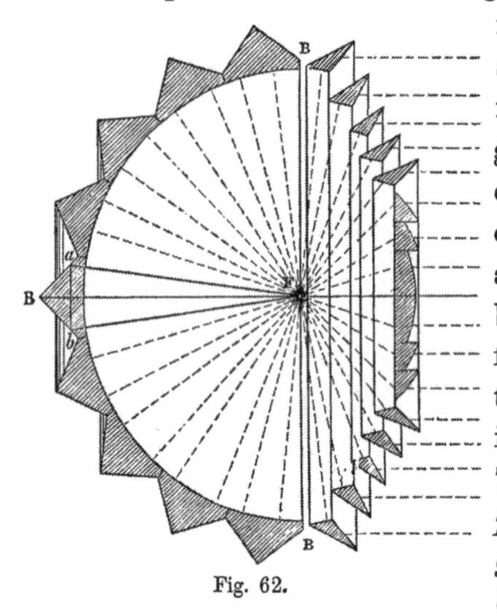

Fig. 62.

This combination should therefore produce *the light of maximum intensity*, and is consequently, so far as arrangement goes, both geometrically and physically perfect; but still, with an oil flame of the ordinary size, some of the rays reflected from the *upper part* of the spherical mirror will not clear the burner, and will therefore be lost. The first double reflecting prism was successfully constructed in 1850 by Mr. J. Adie of Edinburgh, and those of the improved form proposed by Mr. Chance (which will be noticed at its proper date) were made in the most perfect manner by Messrs. Chance of Birmingham for Messrs. Stevenson, in 1861. To a person who is unacquainted with optical principles, the action of these double reflecting prisms is somewhat surprising; for, when they are placed at the proper focal distance between the most dazzling flame and the observer, no trace of light can be seen, although the intervening barrier is only a screen of transparent glass. But if the flame be placed exfocally, or be of too great magnitude, so as to exceed the limit assigned by the formula, the exfocal rays will, instead of being reflected by the mirror, be transmitted freely through it, and therefore lost.

In order to ensure that the light shall be totally reflected at A and I (Fig. 63), which are the points where the angle

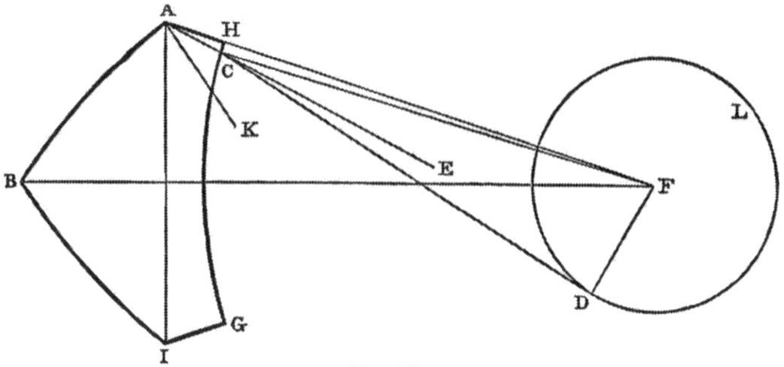

Fig. 63.

of internal incidence is least, Professor Swan, who kindly gave me much assistance in carrying out this design, has given the formula—

$$45^\circ - \frac{\psi}{4} > \sin^{-1}\frac{1}{\mu} + \sin^{-1}\frac{f}{\mu\, d}$$

Putting C F = d, F D = f, μ = the index of refraction of the glass for the extreme red rays of the spectrum, and ψ the greatest admissible value of the angle subtended at the centre of the flame by the face of the zone.

For the radius of the circular arc, which in practice replaces the parabolic arc A B or B I, he obtained the value—

$$r = \frac{d \sin \frac{\psi}{2}}{2 \sin \left(45^\circ - \frac{\psi}{8}\right) \sin \frac{\psi}{8}}$$

And for the co-ordinates of the centre of curvature—

$$a = -d \cos \frac{\psi}{2} + r \sin \left(45^\circ - \frac{\psi}{4}\right)$$
$$b = -r \sin 45^\circ$$

6. *Dioptric Holophotal Revolving Light,* 1850.—*Application of total Reflection to Revolving Lights.*—Sections of the front half of the dioptric holophote (Fig. 62) are equally applicable to a revolving light where there is a central burner, which is for that purpose enclosed by a glass cage made up of those sections. This form of revolving light is shown in Fig. 64, and is geometrically and physically perfect, because it intercepts all the rays by single agents, and these are glass. The central lenses, L L in the original design of 1850 (Fig. 64), instead of being of a rectangular form are circular, so that the reflecting and refracting zones are concentric at their junctions, by which some light is saved. When the light is not required to show all

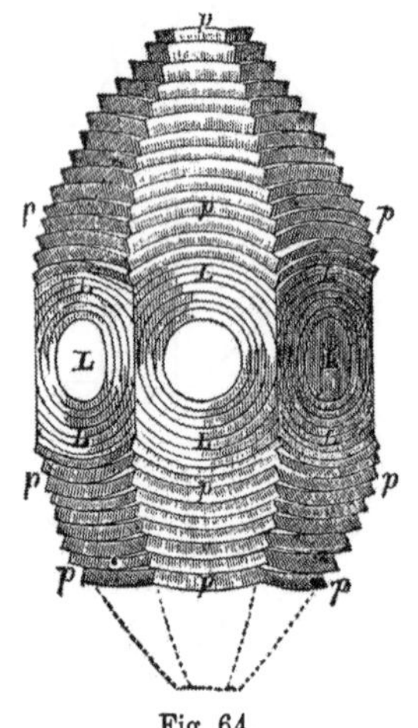

Fig. 64.

round the horizon, a portion of a dioptric spherical mirror may be placed behind the flame on the land side, and the rays reflected by it used for strengthening the opposite seaward arc.

Mr. J. T. Chance, in referring to the relative advantages of Fresnel's revolving light and of the holophotal, says at p. 14 of his paper on optical apparatus:—"In Fresnel's revolving apparatus, as the focal distance of the accessory lenses is less than one half of the shortest focal distance in the system of reflecting zones, the intensity of the light issuing from the former would be scarcely more[1] than one-fourth of that transmitted by the latter; and in addition to this cause of inferiority is the loss arising at the mirrors; so that, on the whole, the modern plan (holophotal) must give light *five or six times* more intense than that of the former (Fresnel's) arrangement."

The holophotal revolving light is now in universal use. It was first introduced in 1850 by Messrs. Stevenson on the small scale, in connection with parabolic reflectors, at Horsburgh Rock, near Singapore, *which was the first lighthouse in which total reflection was applied to revolving apparatus;* and on the largest scale for North Ronaldshay, in Orkney (Plate XXIII.) The Horsburgh prisms were completed in 1850 by Mr. Adie of Edinburgh, and those for North Ronaldshay were most successfully and skilfully made in 1851 by M. Letourneau, the well-known manufacturer of Paris.

7. *Holophotal Fixed Light, varied by Flashes.*—In this arrangement (Roy. Scot. Soc. Arts, 1850) the double agents used in Fresnel's design (No. **18**), and also in Letourneau's

[1] "The words *scarcely more* are used in order to allow for the greater loss of light caused by the prisms than by the lenses, in consequence of the longer paths of the rays in glass."

(No. **27**), are altogether dispensed with, and the whole effect is produced by the single agency of alternate panels of fixed light apparatus $p' p'$, with cylindric refractors L′ L′, and holophotal apparatus p p, both revolving together round a central lamp, as shown in Fig. 65. There will in this, as in Fresnel's form, be a dark interval before the full flash comes round.

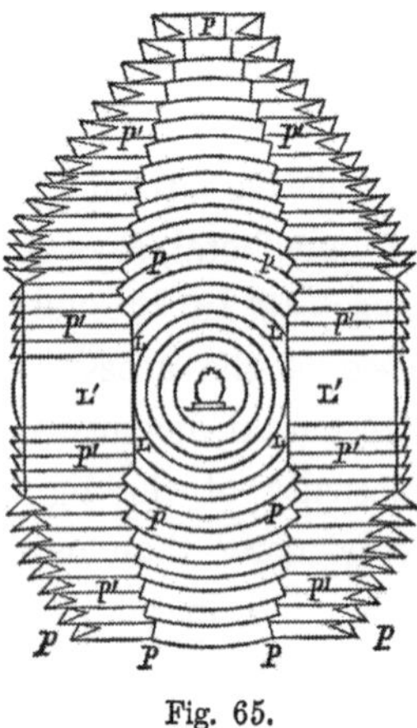

Fig. 65.

The reader will have observed that while perfection of design was attained in Fresnel's *fixed* apparatus, and also in the holophotal *revolving* apparatus described in No. 6, a physical defect attaches to every form of holophote that has to compress the light into *a single beam,* where it is necessary to send the rays back through the flame. The defect is one which is not geometric, and therefore non-existent, where the luminant is a mathematical point, and perhaps hardly appreciable where the radiant is so small as the electric light. The existence of this defect is wholly due to the burner, that indispensable adjunct of an oil flame, through which the oil and supply of air are obtained; and the obstruction is, of course, limited to those rays only which are reflected near the top of the mirror. This explanation will prepare the reader for the no doubt unlooked-for and apparently retrograde step of restoring metallic agency in the design next to be described, which possesses, however, the far more than equal advantage of preventing loss of light by the obstruction produced by the burner.

8. *Improved Catadioptric Holophote,* 1864.[1]—In Figs.

[1] Stevenson's Design and Construction of Harbours, Lond. 1864, p. 251.

66 and 67, *a b c* is the front half of a holophote, which

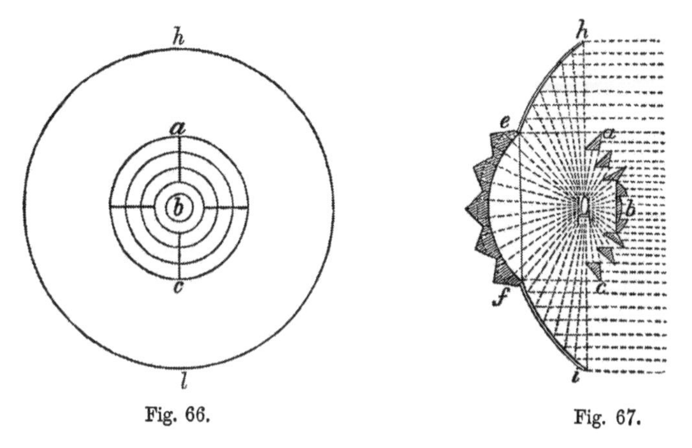

Fig. 66. Fig. 67.

parallelises one half of the light; the other half is intercepted by a portion of a paraboloid, *h e f i*, and a portion of dioptric spherical mirror *e f*, whose action is limited to rays so near the horizontal axis that none fall on the burner and are lost. All that is wanted to make this instrument faultless is to substitute glass for the remaining portion of metallic reflector (*e h f i*) which is used, and this is accomplished in the design next to be described.

9. *Back Prisms.*—The maximum possible deviation of light by means of the Fresnel reflecting prisms of crown glass is limited to about 90°, beyond which the critical angle would be overpassed, and the rays transmitted through the prisms, instead of being reflected by them. In 1867 Mr. Brebner and myself designed what we termed "back prisms," by means of which, rays may be made to deviate from their original direction for about 130°, so that by their use the lighthouse engineer becomes virtually independent of the critical angle. I communicated the description of these prisms to the Royal Scottish

Society of Arts on 6th December 1867. Professor Swan of St. Andrews also independently proposed the same form of prism, a description of which he communicated to the same Society on 9th December 1867, accompanied by general formulæ for its construction. In this form of prism (Fig. 68) the ray *a b* is refracted at *b*, totally reflected at *c*, and again refracted at *d*, so as to pass out parallel to the horizontal axis. Prisms of this kind may be formed by the revolution of their generating section round either a vertical

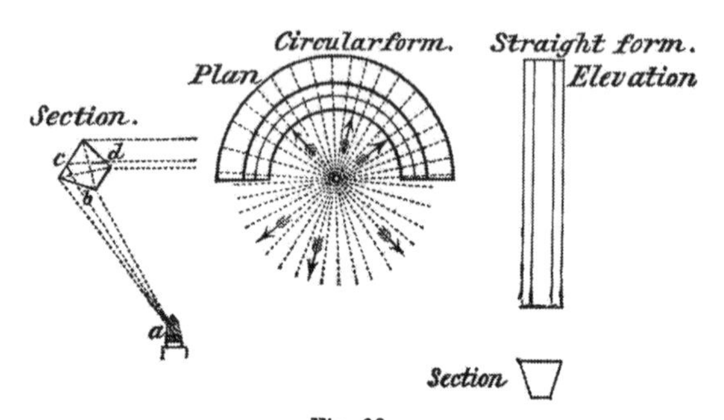

Fig. 68.

or horizontal axis, or may be made straight, as shown in elevation and section in Fig. 68. These prisms were first used at Lochindaal Lighthouse in Islay, and were made by Messrs. Chance in accordance with Professor Swan's formulæ, which will be found in the Appendix.

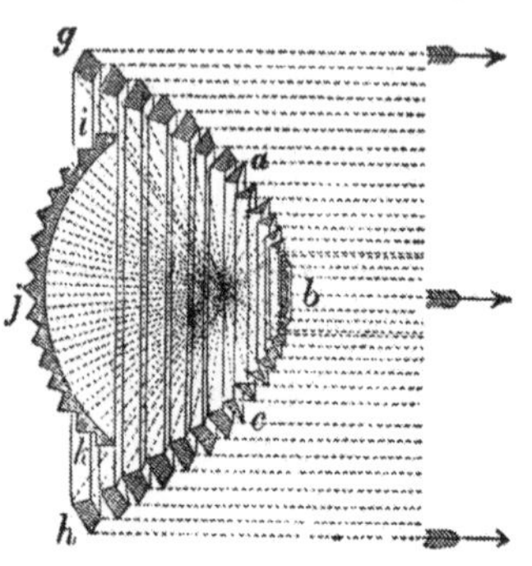

Fig. 69.

10. *Improved Dioptric Holophote* (Fig. 69).— By combining the back prisms, *g a*, *h c*, just described, with a semi-holophote *a b c* subtending

180°, and a portion of the dioptric spherical mirror *i j k*, the whole rays are parallelised, and none of them are lost on the burner, so that this apparatus, being all of glass, is both geometrically and physically perfect.

31. *Mr. J. T. Chance's Improvements of* 1862 *on Stevenson's Dioptric Spherical Mirror.*—Mr. Chance proposed (Fig. 70) to generate the prisms of the spherical mirror round a vertical instead of a horizontal axis, and also to arrange them in segments. He says (Min. Inst. Civ. Eng., vol. xxvi.) :—" The plan of generating the zones round the vertical axis was introduced by the author, who adopted it in the first complete catadioptric mirror which was made, and was shown in the Exhibition of 1862 by the Commissioners of Northern

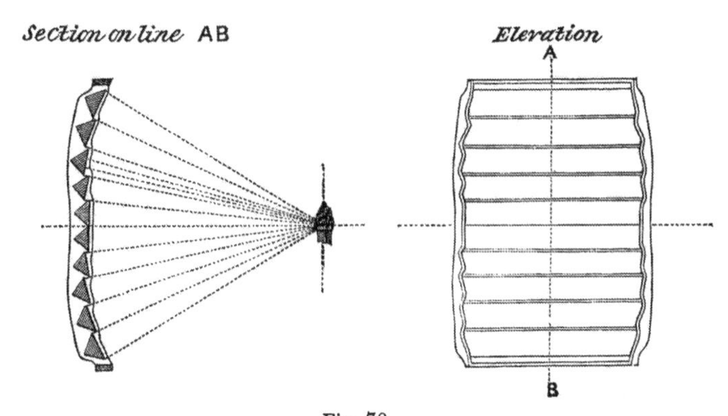

Fig. 70.

Lighthouses, for whom it was constructed, in order to further the realising of what Mr. Thomas Stevenson had ingeniously suggested about twelve years previously."

" During the progress of this instrument the idea occurred to the author of separating the zones, and also of dividing them into segments, like the ordinary reflecting zones of a dioptric light; by this means it became practicable to increase considerably the radius of the mirror, and thereby to render it applicable to the largest sea light, without overstepping the limits of the angular breadths of the zones, and yet without being compelled to resort to glass of high refractive power."[1]

There can be no doubt of the advantage of these improvements, and it is without any intention of derogating from Mr. Chance's merit in the matter that it is added that my first idea was also to generate the prisms round a vertical axis. But the flint-glass which was necessary for so small a mirror could not be obtained in large pots, and had to be taken out in very small quantities on the end of a rod and pressed down into the mould. I was therefore obliged to reduce the diameter of the rings as much as possible; and it was thought by those whom I consulted at the time (Mr. John Adie, Mr. Alan Stevenson, and Professor Swan), that by adopting the horizontal axis the most important and most useful parts of the instrument near the axis would be more easily executed, inasmuch as those prisms were of very much smaller diameter. Mr. Chance not only adopted the better form, but added the important improvement of separating the prisms and arranging them in segments. It would be better, however, if the prisms were constructed with a shoulder, as in the original design, A H, I G (Fig. 63), instead of a sharp edge, as in section, Fig. 70, for the inner concave surface interferes with the passage of the rays reflected from the parts of the sides adjacent to it.

32. *Professor Swan's Optical Agents.*—In the Trans.

[1] Mr. D. Henderson states that M. Masselin and he suggested this arrangement in 1860, but, however this may be, Mr. Chance's was the first publication of the improvement.

Roy. Scot. Soc. of Arts (1867-8) Professor Swan proposed a number of ingenious arrangements and new forms of prisms. Among these is what he termed the Triesoptric prism, by which the rays would undergo two refractions and three reflections. Its execution presents great difficulties in construction, which Professor Swan suggests might be overcome by making it in two pieces, which would be afterwards cemented together by Canada balsam. Among other combinations, for which the reader is referred to the paper itself, is a valuable suggestion for separating the front prisms, and sending light parallelised by prisms placed behind, through the interspaces left between the front prisms. In Fig. 71, *a* are the front, and *b* the triesoptric prisms. The two upper and lower prisms, *c a*, are constructed of flint glass, having a higher index of refraction. It will be observed from the drawing that this ingenious arrangement is nevertheless open to objection, for cones of rays of 30° in front and of 65° at the back are lost when an oil lamp is used. Professor Swan suggests that the greater part of this loss might be prevented by employing the electric light, and placing its lamp horizontally instead of vertically; but in the case of fixed lights showing all round, there would still be a very considerable loss from want of interception, as well as great divergence from the short focal distance of the central back prisms. In the dioptric spherical mirror this evil is avoided, as

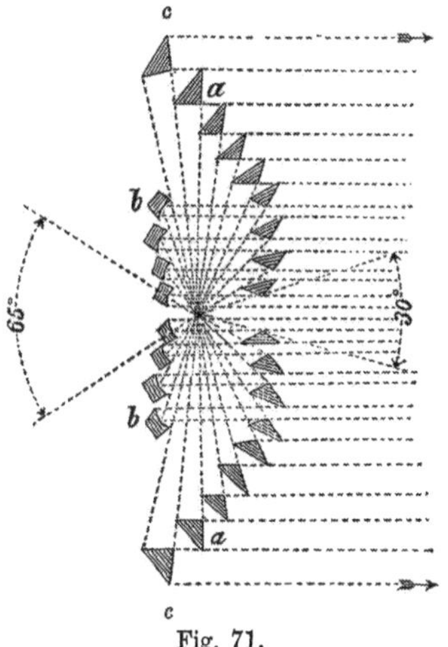

Fig. 71.

every point in its surface is removed sufficiently far from the flame.

We have now traced the gradual progress of improvement from the rudest to the most perfect forms of apparatus for illuminating the horizon with light of equal power, both constantly and periodically.

The FIXED LIGHT *was shown* (**16**) *to have been perfected by Fresnel by means of his two new optical agents, including the first introduction of the principle of total reflection to this kind of light.*

The REVOLVING LIGHT *was also shown* (Fig. 64) *to have been, by single agency, perfected by the application of the holophotal principle, by which total reflection was first applied to moving apparatus, and by which, as shown in* Fig. 69, *all the rays of light diverging from a flame may be condensed into a single beam of parallel rays, with the minimum number of agents.*

CHAPTER III.

AZIMUTHAL CONDENSING SYSTEM FOR DISTRIBUTING THE LIGHT UNEQUALLY IN DIFFERENT DIRECTIONS EITHER CONSTANTLY OR PERIODICALLY.

1. WE have now to consider certain requirements for the ultimate destination of the rays, of a different and less simple nature than those which were dealt with in the former Chapter. Previous to 1855, lighthouse apparatus, having the same illuminating power in every azimuth, was used not only at places where the distances from which the light could be seen were everywhere equal, and where the employment of such apparatus was therefore quite legitimate, but also for places having a sea range much greater in some directions than in others. This indiscriminate application of apparatus of equal power to the illumination of our coasts necessarily involved a violation of economic principle, for the light was either too weak in one direction or else unnecessarily strong in another. At stations where there were several lamps, each with its own reflector, the evil could no doubt be obviated to a certain extent by employing a greater number of such instruments to show in the direction of the longer, than of the shorter range. But from what has been already explained in the previous Chapter, the reader will see that by this method the light could not be distributed *uniformly* over any given sector. In other cases, where perhaps only half of the horizon had to be lighted, a single flame in the focus of a fixed apparatus could also be strengthened by a hemispheric reflector placed on the side next the land, so as to send back the rays, and thus increase the power in the seaward arc, but no attempt was ever made *to allocate this auxiliary light in proportion to the varying lengths of the different ranges, and the amplitudes of the arcs to be illuminated; nor, where a light had to show all round the horizon, to weaken its intensity in one arc, and with the rays so abstracted, to strengthen some other arc, which, from its range being longer, required to be of greater power.* As none of the agents or combinations which we have as yet described were sufficient for dealing with this branch of lighthouse optics, I found it necessary to devise eight new agents, possessing special optical properties, for distributing the rays not *equally* but *equitably.* These new agents, and their combinations with some of those formerly in use, which are required for dealing with special cases, will be described.

The most important of the problems which have to be solved are the following, and, from the principles involved in them, the modes of dealing with different modifications of such requirements can easily be deduced, and examples will be given in which some of these modifications have been carried out.

2. Requirements for Condensing Fixed Lights.

1*st, Where a light has to be seen constantly over only* ONE ARC *of the horizon, the apparatus must be made to compress all the rays from the flame within that one arc, whatever its amplitude may be, and to spread them uniformly over it.*

2*d, Where a light has to illuminate constantly the* WHOLE HORIZON, BUT HAS TO BE SEEN AT GREATER DISTANCES OVER

SOME PARTS OF THE SEA THAN OVER OTHERS, *the apparatus must be made to abstract as much light as can be spared from the shorter ranges, and divert it to the illumination of the longer, so as to allocate all the rays in the compound ratio of the number of degrees, and the squares of the distances from which the light has to be seen in each arc. And the light, which has thus been diverted from any arc in order to strengthen another, must be spread uniformly over the one that has to be strengthened. By this mode of abstraction and addition, there is therefore produced a constant equitable distribution over the whole horizon of all the rays from any single flame.*

3. REQUIREMENTS FOR CONDENSING REVOLVING LIGHTS.

1*st*, *Where a light has to give its flashes periodically over only* ONE ARC OF THE HORIZON, *the apparatus must collect all the rays and send them out periodically in solid beams of equal power over that one arc only.*

2*d*, *Where a light has to illuminate periodically* THE WHOLE HORIZON, BUT WHERE ITS FLASHES HAVE TO BE SEEN AT GREATER DISTANCES OVER SOME PARTS OF THE SEA THAN OVER OTHERS, *the apparatus must be made (as in No.* 2 *of the condensing fixed light) to vary proportionally the power of the flashes whenever they pass over those parts of the sea where the ranges are of different lengths, so as to produce an equitable periodic distribution of all the rays over the whole horizon.*

4. According to the hypothesis which has been generally advanced as probable, the loss by absorption in passing through the air increases in a geometric ratio. The subject, however, is one of which very little is known experimentally, and Sir John Herschell remarks that it proceeds on the *supposition* "that the rays in the act of traversing one stratum of a medium acquire no additional facility to penetrate the remainder." If the absorption were neglected, condensing apparatus, when more than one azimuthal angle has to be strengthened, should, as we have said, be so calculated as to distribute the rays in the compound ratio of the squares of the distances and the number of degrees in each arc. Thus, if n be the number of degrees in an arc to be illuminated, and d the distance in miles to be traversed by the light, then, neglecting atmospheric absorption, the quantity of light to be apportioned to that arc will be proportional to $n\,d^2$. But, if we take account of atmospheric absorption, supposing q to be the quantity out of a unit of transmitted light which escapes absorption after passing through a mile of air, then the whole light needed by the arc to be illuminated will be proportional to $m = n\,d^2\,q^d$.

Supposing now that L is the whole 360° of available light emitted by the lighthouse apparatus, the quantity to be apportioned to the given arc will be

$$\frac{m\,L}{\Sigma\,m};$$

where $\Sigma\,m$ denotes the sum of the several numbers m computed for the respective arcs of the horizon to be illuminated.

5. What follows will show the modes of fulfilling the various conditions which have been enunciated. But before entering upon this description, it is well to notice, in case of any misconception, that the form assumed by the emergent rays of a fixed light on the condensing system, is neither a solid beam of parallel rays like that from an annular lens, nor yet a zone of rays diverging *naturally* in azimuth all round, like that from an ordinary fixed apparatus, but is intermediate between these, being a *solid angle* *or wedge* of light, strengthened by those rays which would naturally diverge in other directions, but which are diverted and spread over the given sector. This is shown perspectively in plan in Fig. 72, in which *L* represents the position of the lighthouse, the radius of the circle *L D*, or *L D′* the range or distance at sea from which the light can be seen, and *D′ L D d* the solid horizontal angle that is to be illuminated, and into which all the light *D A D′*, which would naturally diverge over the rest of the circle, must be compressed, and over which it must be uniformly spread.

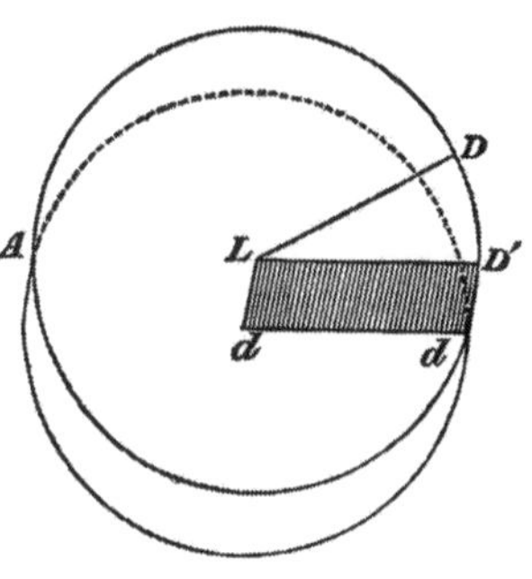

Fig. 72.

The condensing system divides itself into the following different heads :—

FIXED LIGHTS.

1st, Which compress all the rays from the flame into a *single sector* in azimuth.

2d, Which compress all the rays into *more than one sector* of unequal range.

3d, Which illuminate constantly the *whole horizon*, but which vary the strength of the light in the sectors of different range.

REVOLVING LIGHTS.

1st, Which compress all the rays into flashes passing periodically over *a single sector.*

2d, Which compress all the rays into flashes of unequal power, which illuminate periodically *more than one sector* of unequal range.

3d, Which periodically illuminate the *whole horizon,*

but which vary the strength of the flashes whenever they pass over sectors of unequal range.

6. Optical Condensing Agents.

1. *Condensing Straight Prisms*[1] which, either by reflection or refraction or both, cause a ray *f r* (Fig. 73), proceeding in any compass bearing from the main apparatus *A A*, to emerge parallel (in the direction *e g*) to the corresponding ray *f b*, which proceeds in the same compass bearing from another part of the apparatus; and so of any other ray *f c*, which is bent parallel to the ray *f a*.

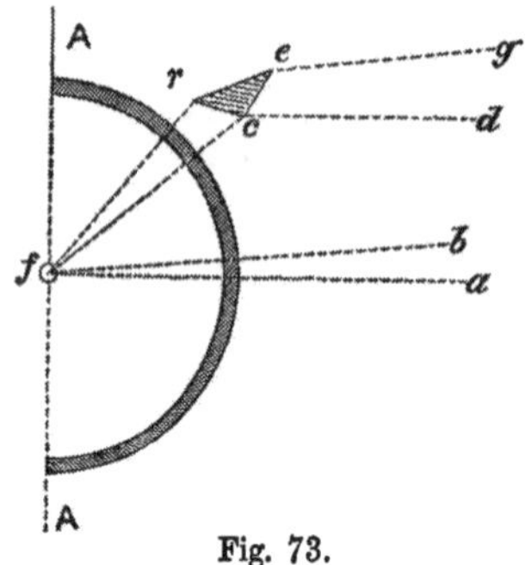

Fig. 73.

2. *Right-angled Expanding Prisms* (introduced at the Tay leading lights in 1866).—These prisms, shown in Fig. 74, are right-angled vertically; while *in plano* two of them C_1 C_2 are semi-rings, and the third C_3 is a semi-cone. Parallel rays in passing vertically upwards in a semi-cylindric beam, and falling normally on the bases *a* of the prisms, enter the glass, are reflected by the sides *b*, and pass out horizontally and normally at the other sides; but as the prisms are bent through a circular segment *d′ a d*, *in plano*, the rays, when they emerge, will be spread

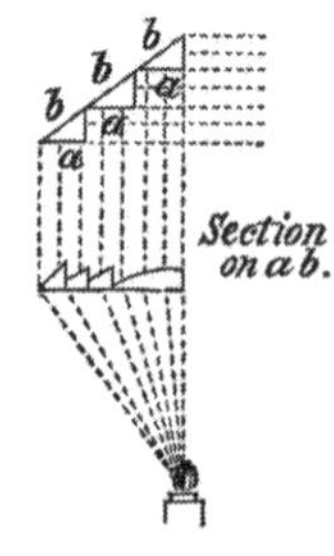

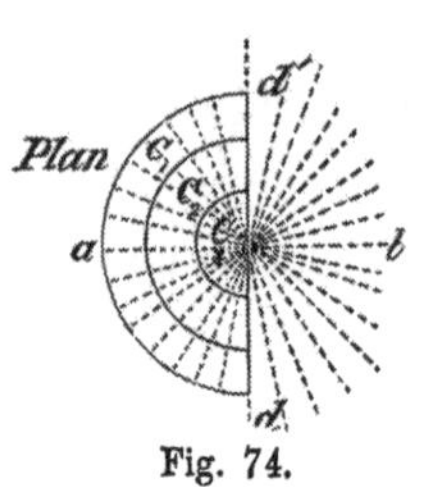

Fig. 74.

[1] On a simple method of distributing naturally diverging rays of light over any azimuthal angle. By T. Stevenson (Edin. New Phil. Journal. April 1855).

over the same angle $d'\,b\,d$ in azimuth by reflection only, for both the immerging and emerging rays enter and leave the prisms without being refracted. Thus, whatever be the horizontal angle subtended by the prisms, the incident parallel rays will be spread over the same angle. As the prisms shown in the figure subtend 180°, the light will in this case be spread over half the horizon, $d'\,b\,d$.

3. *Twin Prisms.*[1]—This kind of prism was designed for carrying out within the smallest possible space the ingenious proposal of Professor Swan (described in Chapter II.), of placing prisms behind others, so as to cause light coming from those behind, to pass through spaces left for the purpose between those in front. The twin prisms (Fig. 75) are formed by cutting out the apex (shown black in diagram) of a straight prism so as to reduce its size and provide more space between it and the next prism for the passage of rays coming from another prism behind. This arrangement not only effects a saving of light by shortening the length of glass traversed by the rays, but it also admits of a very great diminution of the size of the apparatus and lantern.

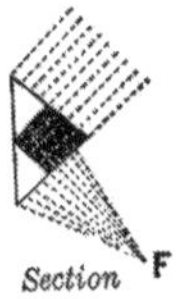

Section F

Fig. 75.

4. *Differential Lens.*—The differential lens which I proposed in 1855 should have no divergence in the vertical plane (excepting that due to the size of the flame), while the horizontal divergence may be adjusted to any required amount, by simply varying the radius of curvature. One face of this lens (Figs. 76, 77, and 78) remains the same as in the ordinary annular lens, while the other face, though straight in the vertical plane, is ground to different curves in the horizontal, so as to produce the required divergence.

[1] On an improved optical arrangement for azimuthal condensing apparatus for Lighthouses. By Thomas Stevenson (Nature, 26th August 1875).

Fig. 76 shows in front elevation the outer, and Fig. 77 the inner face; while Fig. 78 gives the middle horizontal section

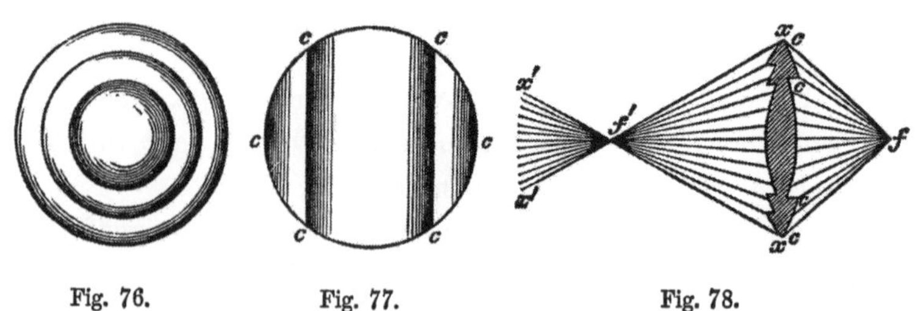

Fig. 76. Fig. 77. Fig. 78.

in which $c\ f\ c$ are rays coming from the flame, which, after passing through the lens x, converge to the vertical focal plane f', and afterwards diverge through the smaller *horizontal* angle $x'\ f'\ x'$. When the angle $x'\ f'\ x'$ over which the light has to be seen is, as in this case, smaller than the angle $x\ f\ x$ which the lens subtends from the flame, the radius of convexity of the inner face must be greater than that of the outer face, and *vice versa.*

5. *Differential Refractor.*—Is the application to the refractor of the same principle which has been described above for the lens.

6. *Differential Reflector of single agency.*—It is possible to condense the rays over a horizontal sector by a single agent. For this purpose the vertical section must be parabolic, while in the horizontal it must be of such hyperbolic, elliptic, or other curve, as will most advantageously give, in each case, the required horizontal divergence. The difficulty in construction might be most easily overcome by using the old form of mirror in which reflection is produced by small facets of silvered glass, bent or ground vertically and horizontally to the parabolic and elliptic curves. If the edges of the facets were cemented together with Canada

balsam, the large loss of light which takes place at the edges of each facet in the old reflectors will in great measure be saved. There will not, as formerly, be any refraction of the rays in passing through the edges of the facets, and thus the whole mirror will become practically *monodioptric*, or, in other words, optically nearly the same as if it had been made of one whole sheet of glass. It would be a further improvement to select for the foci of the facets different points in the flame, so as to secure the accurate destination of the brightest sections of the flame. When coloured lights are wanted the facets would consist of glass tinted with the required colour, so as to render stained glass chimneys unnecessary, thus saving light. Professor Tait was kind enough to investigate the mathematical conditions of the differential mirror, and in the Proceedings of the Royal Society of Edinburgh of 1871 he gives, by a quaternion integration, the formulæ for its construction. An investigation by the ordinary methods will be found in the Appendix in more detail.

The difficulty of construction has hitherto prevented the employment of this instrument. One mirror was made in accordance with the formulæ; but the surface was imperfectly formed and the light was not very accurately directed.

7. *Condensing Catoptric Spherical Mirror.*—Plate XVI. represents different modes of condensing light by means of a spherical mirror which has a sector cut out opposite to the arc that may be weakened. The rays which thus escape are received (Fig. 1) on an ellipsoidal mirror e e, which converges them to its other conjugate focus F, while an interposed plane mirror *m m*, deflects them through a smaller aperture in the spherical mirror, opposite to the arc

that requires to be strengthened. This aperture may be reduced to still smaller dimensions by a lens L (Fig. 2). It is, of course, desirable to diminish as much as possible the size of this opening, because the condensation of the rays depends on the relative sizes of the two apertures. If they were equal, as many rays would pass in the one direction as in the other, and no condensation would be effected. The same result may be obtained by means of paraboloids, having different focal lengths, and Plate XXXII., Fig. 2, shows a similar design, in which an ellipsoid A B, and hyperboloid C D, are used.

8. *Spherical Mirror of unequal area.*—A very simple mode of allocating the light to the different sectors would be to place between the lamp and the apparatus a spherical mirror varying in height inversely with the lengths of range in the different azimuths, as shown in Plate XV. That portion of the light which can be spared in one direction is thus made to assist in another where the range is greater, while the remainder of the rays passing above the mirror at the places where it has been lowered, are allowed to continue in their original directions. The mirror should therefore be so cut down that its different heights will represent inversely the varying distances of the neighbouring land from the lighthouse.

7. Fixed Condensing Lights for a single Sector.

1. The first design[1] which I made for this purpose is shown in section and elevation (Figs. 79, 80), where light, parallelised into a single beam by a holophote *a b a*, is made to converge in azimuth to a vertical focal line *f'* by the straight

[1] Edin. New Phil. Journal, 1855.

refractor *c c*, after which it diverges over any horizontal sector G f' H. This design satisfies the conditions of the problem so long as it is not essential to spread the light *uniformly* over the sector G f' H. But in almost every case uniformity

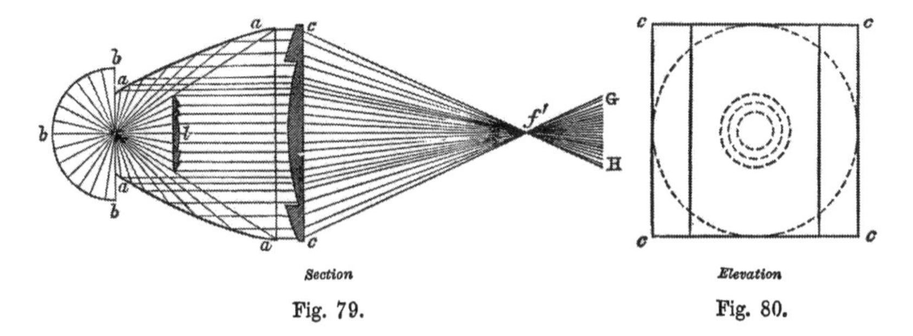

Fig. 79. Fig. 80.

of distribution is essential. To provide for this, a num-

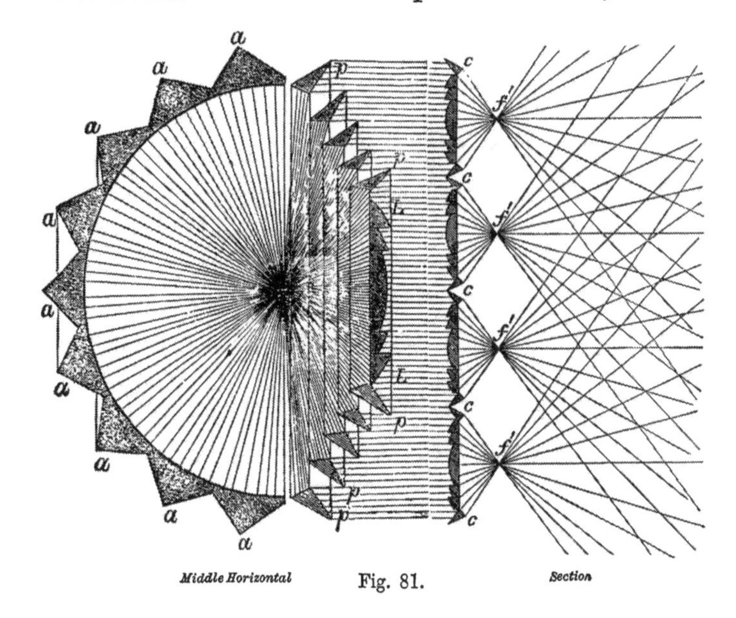

Fig. 81.

ber of independent refractors, *c c* (Fig. 81), are substituted for the single refractor. Each of these refractors spreads all the rays incident upon it over the required sector, which is thus illuminated equally throughout. These were the first, and are indeed the fundamental designs for condensing and spreading all the light over any sector. But the same result may be effected in other ways, examples of which will next be described.

2. *Condensing Quadrant.*[1]—This combination (Fig. 82) consists of a central or main apparatus of the ordinary kind, $b\ b\ b$, with auxiliary straight prisms p and p' (**6**, 1) so as to intercept on each side 45° of the light from the main apparatus, and cause the different rays to emerge parallel to other rays coming from the unobstructed central quadrant of

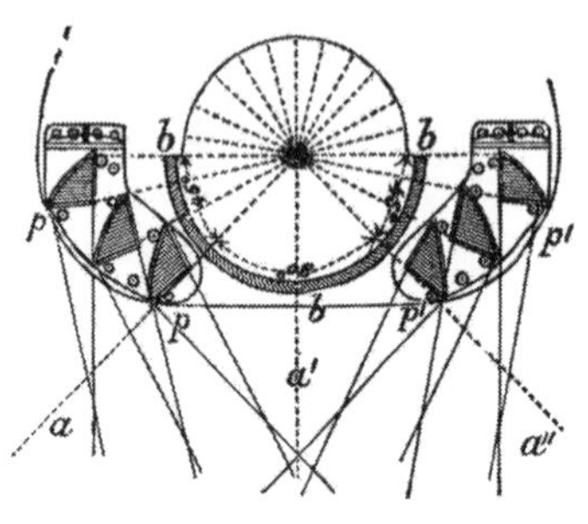

Fig. 82.

the main apparatus. Those prisms marked p' send out rays in the same compass bearings as those from the arc $a\ a'$; while those marked p send their rays parallel to those from the arc $a'\ a''$. The straight prisms are in this case concave on the reflecting side, and straight on the refracting sides. The whole light will therefore be condensed equally over 90°.

3. *Condensing Octant* (Tay Leading Lights, 1866).—This apparatus was remarkable, at the time it was made, from its containing *every kind of* dioptric agent then known, viz. the lens, reflecting prisms, and cylindric refractor of Fresnel (Chap. II.); and the holophotal, the condensing, the right-angled expanding, and the double reflecting prisms (Chap. II.) The apparatus

[1] Holophotal System, London, 1859.

is shown in vertical and horizontal sections (Figs. 83, 84), and in front elevation (Plate XX.) The main apparatus is a half-circle of a fixed light instrument, *a b c* (Fig. 83), and *b b b*

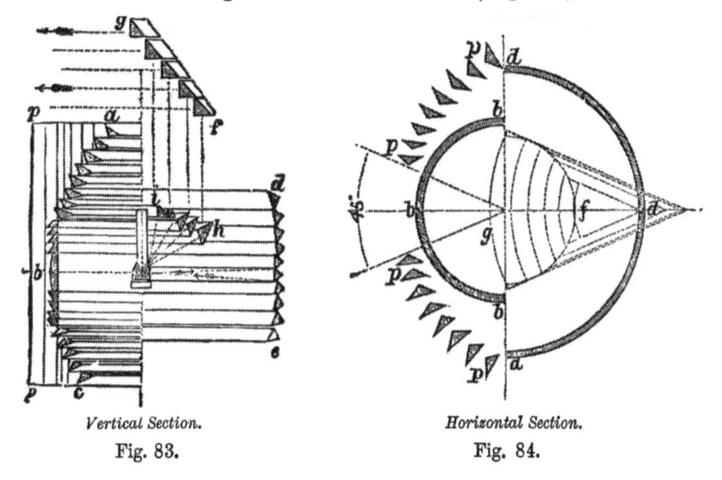

Vertical Section.
Fig. 83.

Horizontal Section.
Fig. 84.

(Fig. 84), 45° of which spread the light *directly* by single agency over the required arc, *p g p* (Fig. 84). On either side of this middle arc are placed in front of the main apparatus, straight condensing prisms, *p p p p*, which also spread over 45° all the light which falls on them after leaving the main apparatus. In this way the whole of the *front* hemisphere of rays is parallelised in the vertical plane, and spread equally over 45° in azimuth. Let us next consider how the hemisphere of back rays is also to be condensed into the same seaward arc. These rays are received in part by the dioptric spherical mirror *d d a* Fig. 84, and *d d* Fig. 83, and are returned through the flame, where, mingling with the front rays, they are finally distributed over the 45° which are to be lighted. The light which passes above the spherical mirror is parallelised by a half-holophote, *i h* (Fig. 83) (which is fixed above the flame), and bent by it vertically upwards, when, falling on *g f*, the expanding prisms (**6,** 2), it is finally spread over the required arc.

Thus the whole light is condensed and distributed with strict equality over the 45° by means of six optical agents, involving in no case more than four refractions and four total reflections. For the upper double agents, the back prisms afterwards invented and already described (Chap. II.) might be substituted. This apparatus, which was manufactured by Messrs. Chance in the most perfect manner, was erected at the mouth of the Tay, and a duplicate was exhibited by the Commissioners of Northern Lights at the Paris Exhibition of 1867, and is now placed in the Edinburgh Industrial Museum. The only other case in which the expanding prisms have as yet been employed, is at Souter

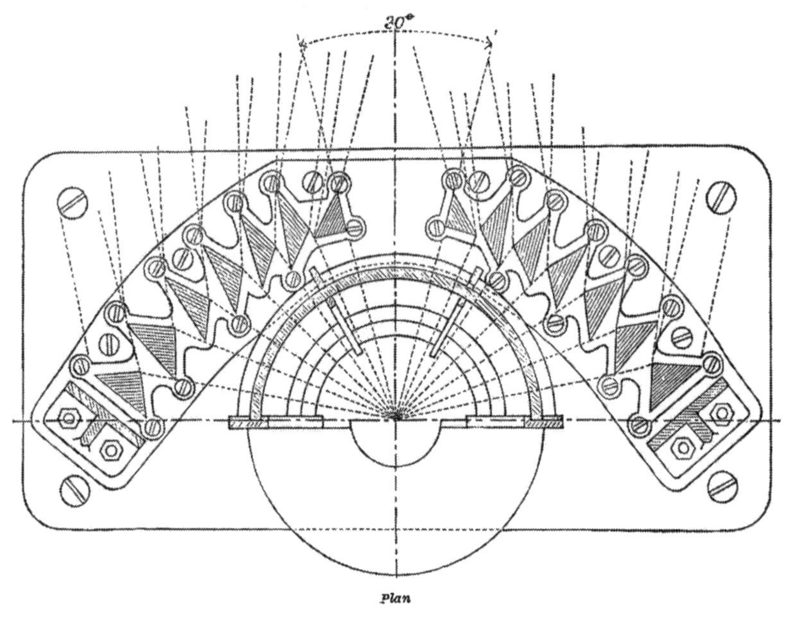

Fig. 85.

Point electric light, the apparatus of which was designed by Mr. J. T. Chance, for the Trinity House, in 1870 (Fig. 4, Plate XXXI.) In this design Mr. Chance employed these prisms for carrying out a suggestion of Mr. Douglass, to send a portion of the light from the principal apparatus, downwards from the lantern to the bottom of the tower, where the expanding prisms distribute the rays over a danger lying at some distance from the shore.

4. *Condensing apparatus for $\frac{1}{12}$th of the circle.*—Figs.

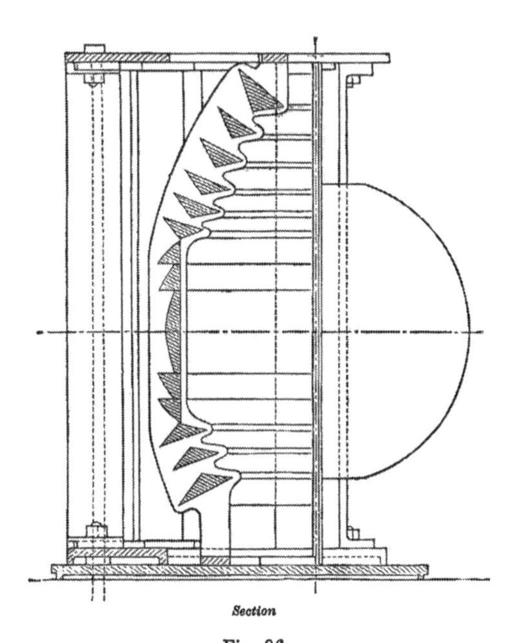

Fig. 86.

85 and 86 represent the condensing light of Cape Maria Van Diemen, New Zealand, in which the whole light is condensed into a sector of only 30°, being the greatest compression hitherto effected by the condensing system.

5. *Twin Prism Condenser.*—Fig. 87 shows the apparatus

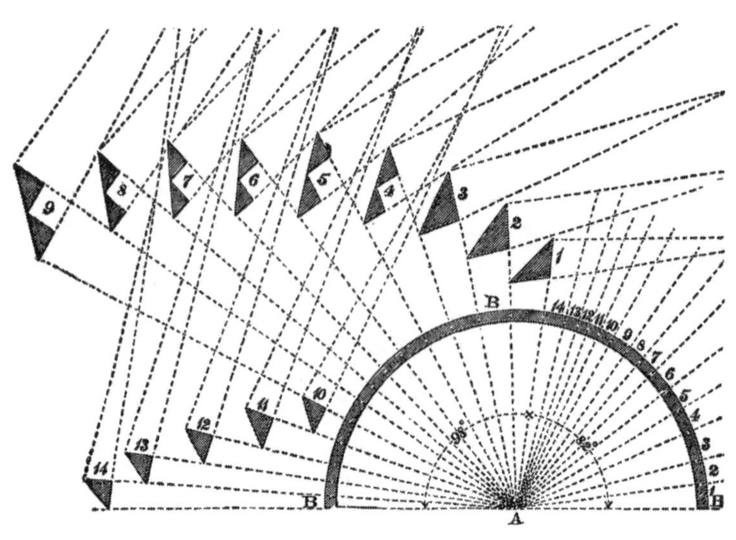

Fig. 87.

for Lamlash, near the island of Arran, in the Firth of Clyde, constructed in 1876, in which the twin prisms (**6**, 3) were first applied. Their action will be easily understood without further description, by the numbers shown on the diagram.

8. Fixed Condensing Lights for more than One Sector of Unequal Range.

1. *Isle Oronsay Apparatus.* (Lighted in 1857.)—Isle Oronsay is situate in the narrow Sound of Skye, and throughout nearly the whole of the illuminated arc of 167°, it does not require to be seen at a greater distance than three or four miles, while in one direction (down the Sound towards A and B, Fig. 88), it can be seen for about fifteen

miles, and in another up the Sound (towards C and D), it can be seen for about seven miles.

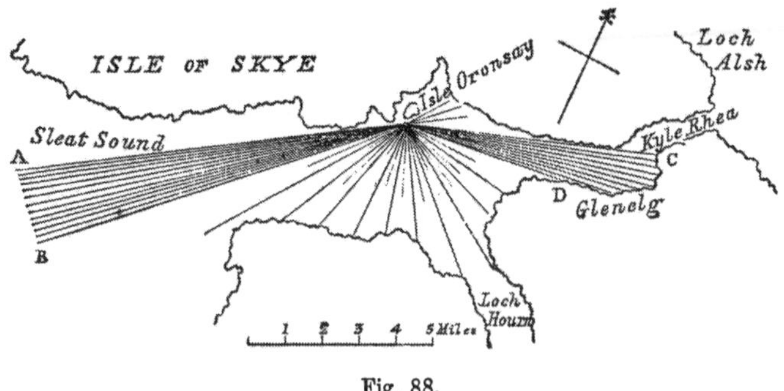

Fig. 88.

The apparatus shown in horizontal section (Fig. 89) was designed in order to reinforce the ordinary fixed light apparatus over the two sectors of greater range. This is effected by distributing in those directions the spare light of 193°, which would otherwise have been lost on the landward side, or, if returned through the focus in the ordinary manner by means of a spherical mirror, would have equally strengthened that portion of the light which is already sufficiently powerful. The formula for condensing the light unequally for different arcs and distances has already been given (4). But the magnitude of the apparatus on which the visual angle depends, forms an element of some importance, especially in narrow seas, such as the Sound of Sleat. It is obvious, however, that the influence of this element must be circumscribed within certain limits. The Oronsay light was therefore, after due consideration, allocated nearly in the arithmetical ratios of the distances, and some such allocation appears warranted from nautical considerations connected with the locality. Moreover, there are difficulties in construction, and also in connection with the amount of available space in the lantern which had to be taken into account. The arc down the Sound could not have been made small enough without cutting through the central bull's eye of the lens, and this would not have been advisable.

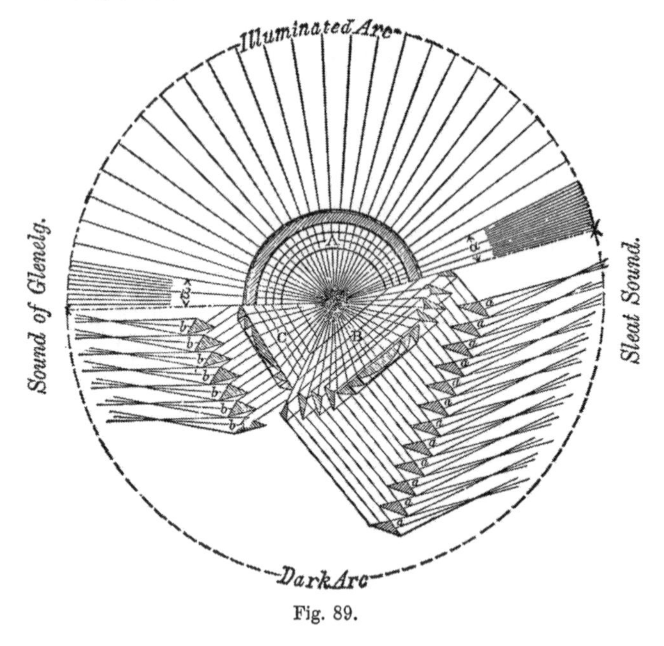

Fig. 89.

In Fig. 89, A is a sector of 167° of a fixed light apparatus. This sector subtends the entire arc A C, in Fig. 88, which is all that has to be illuminated, so that the rest of the light on the landward side, amounting to 193°, is spare light, of which 129° are parallelised by the portion of the holophotal apparatus B, and after falling on a series of twelve equal and similar straight prisms *a*, are again refracted, but in the horizontal plane only; and lastly, after passing through a focal plane (independent for each prism), they emerge in a series of 12 equal wedges, each having a horizontal divergence of about 10°, and in all of which the rays are respectively parallel to those of the directly diverging sector *a*, from the main apparatus, which is therefore strengthened. Each of these supplemental prisms spreads its light over the whole arc A B (Fig. 88), and the portion *a* of the main apparatus does the same *directly* by a single agent. As the light of 139° is in this manner condensed into an arc of about 10°, the power, disregarding the loss in transmission through the prisms, is about *fourteen times that of the unassisted apparatus*, and should doubtless be amply sufficient for a range of 15 miles. In like manner, the light parallelised by the other lens C, is refracted by the prisms *b b*, so as to strengthen the arc *β*, from the main apparatus, which has a range of 7 miles, viz. between C and D (Fig. 88). The greater number of rays which are represented in the chart (Fig. 88) as passing through the arcs A B and C D, is intended to indicate the additional density due to the action of the auxiliary prisms. The condensed light at Isle Oronsay appears to be equal to a first-order apparatus, so that, with a lamp consuming only 170 gallons of oil per annum, a light is obtained (in the only direction in which great power is required), equal in effect to that of a first-order light consuming about 800 gallons per annum.

It may be satisfactory to add the observations which were made on the trial of Isle Oronsay light, which was the first of its kind, by the late Mr. James M. Balfour, who not only took charge of the erection of the apparatus, but to whom is due the execution of the drawings and calculations. In a letter, 17th October 1857, he says, "On Friday I got the apparatus permanently fixed at Isle Oronsay, and at night we steamed out to try the light. More I cannot say than that it was satisfactory beyond our utmost expectations. The prisms throw a light down Sleat Sound superior to any first-class light in the Northern Lights Service, and the light up to Glenelg Bay from the smaller set of prisms is little if at all short of the power of a first-class light, although a large portion of the beam is intercepted by a crossing of astragals * * I have not the slightest hesitation in asserting that the Oronsay light is the best in the Service, both as respects actual power where power is wanted, and also as regards economy of maintenance." Lights, on the same principle, were at the same time erected at Kyle Akin and Runa Gall, and shortly afterwards at Macarthur's Head in Argyllshire, and at several other parts of the coast of Scotland. This kind of condensing light, which is now in general use where oil is used, has also been successfully employed for the electric light of South Foreland, which was designed by Mr. J. T. Chance. The electric light of the Lizards, which was designed by Dr. Hopkinson, is also on the same principle.[1]

2. *Lochindaal Condensing Apparatus.* (Lighted 1860.)—The single acting prisms for operating on the back light (Figs. 68, 69) were first employed at Lochindaal, in the island of

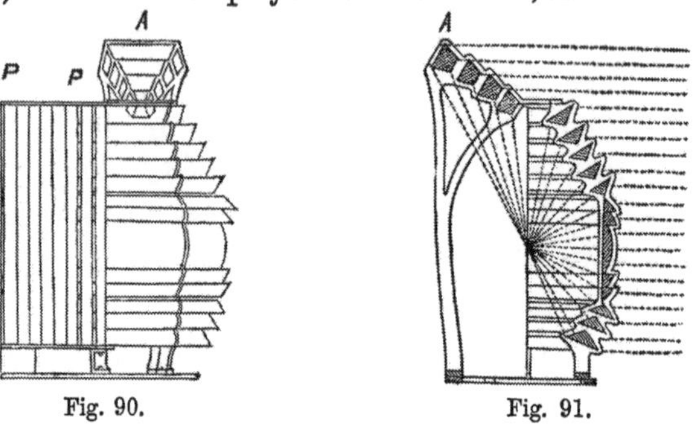

Fig. 90. Fig. 91.

Islay. The apparatus shown in elevation and vertical section

[1] Min. Civ. Eng., vol. lvii.

(Figs. 90, 91) consists of the main fixed light apparatus, surmounted by a panel of "back prisms" A, and supplemented by straight prisms P. The arc of greatest range is strengthened by the "back prisms," which throw the light at once, by single agency, over the arc. These prisms were next used at Stornoway Apparent Light, where they were of a straight form (Chap. V. p. 159). They have also lately been employed by Mr. Chance and Dr. Hopkinson for the electric light.

9. Fixed Condensing Lights of Unequal Range, which constantly illuminate the whole horizon.

The mode of illuminating the whole horizon by means of a fixed apparatus, which will allocate the rays unequally among the different ranges, will easily be understood from the descriptions of the condensing spherical mirror (6, 7), and the spherical mirror of unequal area (6, 8), or as shown in Plate XXXVII.

10. Condensing Apparatus for Steamers' Side Lights.

The condensing principle is equally available for steamers' side lights, for which purpose the apparatus is placed in a small tower in front of the paddle boxes, so as to distribute all the rays with strict equality over 112° 30′, which is the arc prescribed by the Board of Trade regulations. In order to give a steady light, the apparatus and burner are hung on gymbals. This application of the condensing principle was first introduced into the "Pharos," Northern Lights Tender, in 1866 (Plates XVII. and XVIII.) Two of the "Anchor" Line of Transatlantic steamers were also, in

1873, furnished with similar condensing apparatus placed in iron towers, large enough to admit a sailor to trim the lamp, and having means of entrance during bad weather, provided below the deck.

11. Revolving or Intermittent Lights, which condense the light into one sector.

In Fig. 92 a dioptric holophote throws its parallelised light on curved condensing prisms P P, which are so con-

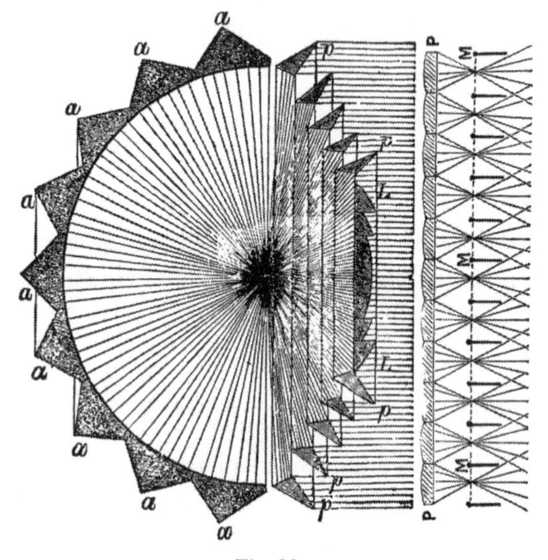

Fig. 92.

structed as to give the required amount of divergence, while masks M, turning horizontally on hinges, cut off the light either slowly or suddenly, as may be wanted for producing a revolving or intermittent light, both of which will thus be made to condense the whole of the rays uniformly over this one sector.

12. Revolving Condensing Lights of Unequal Range, which do not illuminate the whole horizon.

Repeating Light.—Fig. 93 shows an apparatus which I designed in 1860, in which plane mirrors M revolve on an endless chain passing over rollers placed outside of the main apparatus, for altering the direction of the flashes after they pass into the dark arc on the landward side, so as to cause the lenses L L to *repeat* their flashes over the seaward arc which requires strengthening. The only objection which was found on trying this apparatus was, that when viewed at short distances, stray light, reflected from some part of the apparatus, produced a series of irregular coruscations, resembling a miniature display of fireworks, but these fainter flashes might not perhaps injuriously affect the character of the light to a distant observer.

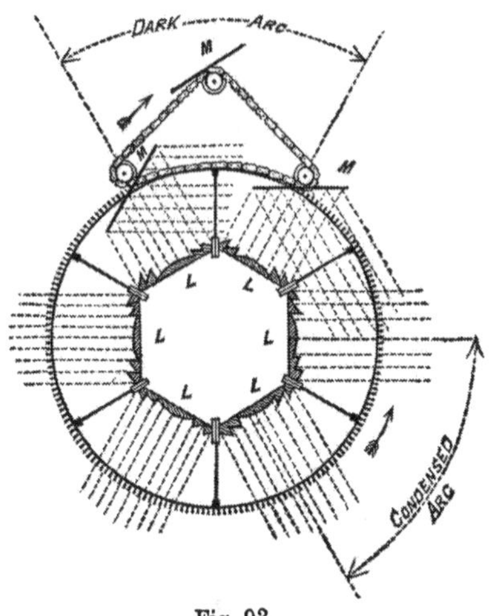

Fig. 93.

The condensing spherical mirror (**6,** 7) and mirror of unequal areas (**6,** 8), which were described for fixed lights, will also be found applicable for revolving lights in which the luminous beams are not required to sweep over the whole horizon.

13. Condensing Revolving Lights which periodically illuminate the whole horizon, but which vary the strength of the Flashes in passing over certain sectors.

The spherical mirror of unequal areas (**6**, 8) is obviously equally well suited for those which revolve. By means of it or the condensing mirror (**6**, 7), any case may be satisfactorily dealt with.

14. Condensing Intermittent Lights.

Although the distinctions about to be described do not strictly fulfil the definition which we have given of the condensing principle, inasmuch as they do not show light of *unequal* intensities in different azimuths, they nevertheless require condensing agents, and will therefore most appropriately be described here.

1. Figs. 94 and 95 show straight refracting or reflecting prisms, which revolve and intercept certain of the rays from a central fixed light apparatus, so as to produce perfect darkness over the sectors they subtend, while they spread the rays which they intercept, uniformly over, and thus strengthen the intermediate sectors, which are illuminated directly by the central apparatus. The peculiar property of this arrangement is that the power is increased in proportion to the duration of the intervening periods of darkness. Thus, neglecting the loss by absorption, etc., the power is *doubled* when the periods of light and darkness are *equal*, *trebled* when the dark periods are *twice* as long as the light, and so on in proportion, while in every case the rays are spread uniformly over each illuminated sector. The con-

densing intermittent apparatus which was first proposed in

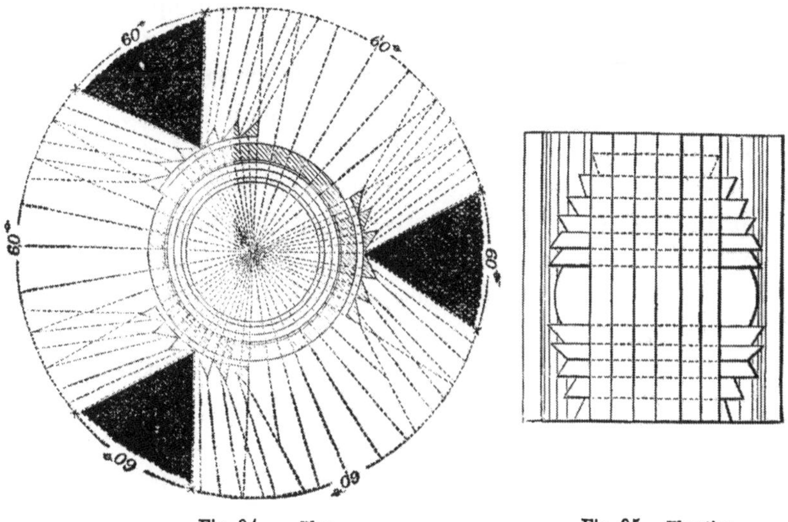

Fig. 94. *Plan* Fig. 95. *Elevation*

1872,[1] was not employed till 1878 at Barrahead, in the Hebrides. The Cabot Island light, in Newfoundland, is also on the same principle, and it is reported that it "affords great satisfaction." The apparatus for these lights was very correctly made by Messrs. Barbier and Fenestre, who exhibited a duplicate at the late Paris International Exhibition, for which they received a medal; and, though not myself an exhibitor, I was also awarded a gold medal. Messrs. Sautter, Lemonnier, et Cie., published a description of apparatus on the same principle in 1879, and they have introduced it at several places.

2. *Condensing Intermittent Lights of unequal periods.*—Plate XIII. shows an example of the apparatus where the continuity of the light is broken up into long and short periods

[1] Royal Scot. Society of Arts, vol. viii.

of darkness alternately; and Plate XIV. shows an example of a similar light, where the continuity is broken up by one long and two short periods of darkness alternately. The light periods in each case are of precisely the same intensity and of equal and definite duration, and are in the first case *two and a third* and in the latter *three* times the strength of the fixed light. In these forms of apparatus the vertical prisms only are made to revolve, and by their adoption two new characteristics are added to the lighthouse system independently of the changes which might be produced by varying the speed of revolution of the prisms. Any existing fixed light can easily therefore be altered in character and increased in power by the simple addition of the condensing prisms. It will be readily seen that these lights are different in character from the group flashing light of Dr. Hopkinson (Chapter IV.), in which the duration of the flashes is dependent *solely on the natural divergence due to the size of the flame.* The drawings will be readily understood without further explanation.

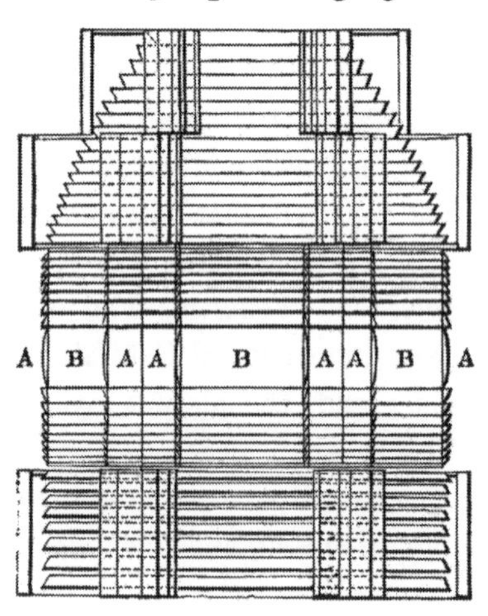

Fig. 96.

3. *Condensing Intermittent Light with Differential Refractors* (Figs. 96, 97).[1]—A still more perfect form of the condensing intermittent light can be produced by availing ourselves of the property of the differential lens (**6**, 4), which, in this case, takes the form of the differential refractor (**6**, 5). This refractor

[1] Min. Ins. of Civ. Eng., vol. lviii. 1878-9.

having its centre of inner curvature at O, Fig. 97, is substi-

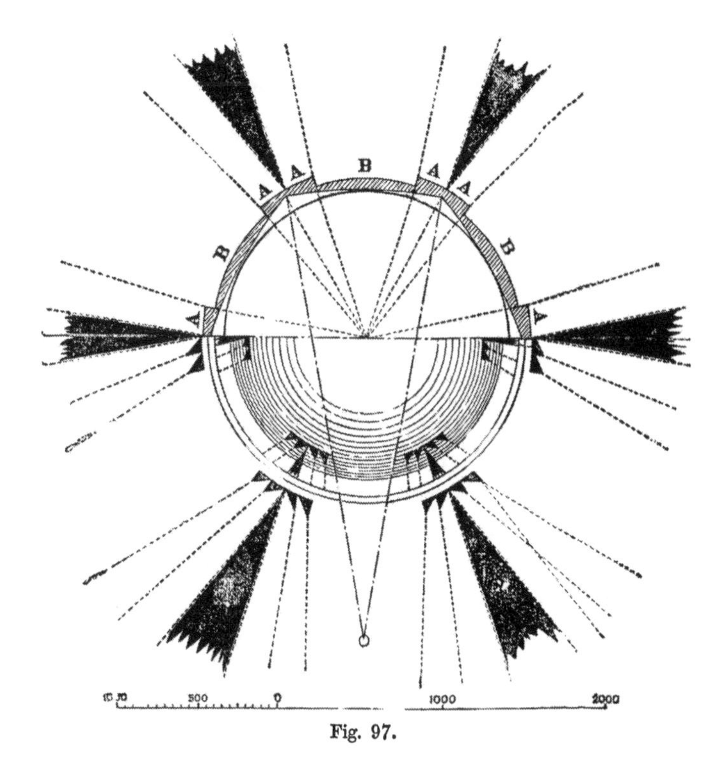

Fig. 97.

tuted for the ordinary refractor of Fresnel, and the central drum, instead of being a continuous hoop, becomes polygonal, A B A (Figs. 96, 97). The design shown (half in plan and half in section through the focus, Fig. 97) is that of the new apparatus at the Mull of Galloway, Wigtonshire, where *all* the parts revolve together. The outside vertical profile of the differential refractor is the same as that of Fresnel's refractor, while the inner is curved horizontally to the radius O A, and the upper and lower joints, in order to avoid intercepting light, are concave conoidal and convex conoidal respectively. By the compound horizontal and vertical action of this *single* agent, the whole intermittent effect is produced, so that double agents are no longer required excepting for the upper and lower reflecting prisms. The joints are also made coincident with the paths of the rays *after refraction*, an improvement which was previously suggested for ordinary apparatus by M. Allard. The straight prisms may be also adjusted so as to counteract the effect of unequal incidence of the rays on each side of the centre of the refractor. The lightkeepers at Point of Ayre, Isle of Man, twenty-one miles distant, report that the new light of Mull of Galloway "appears to us to be nearly double the strength in intensity of what the old reflector light formerly was." Plate XXXIV. shows another design, in which dioptric spherical mirrors are used.

4. *Fixed White Light changing to Fixed Coloured.*—If two sectors of a fixed apparatus (Plate XXXVI.) be covered with coloured glass, and if coloured condensing prisms be placed in front of so much of the intermediate sector as, when spread over the adjoining coloured sectors, will make them equal in intensity to the white light coming from the uncovered part of the intermediate sector, then, if the coloured glass and prisms be made to revolve, the effect will be a fixed light of equal power constantly in view, and changing its character from white to red alternately, without any intervening dark period, or any waxing or waning of the light, and capable of illuminating simultaneously the whole horizon. (See page 132.)

5. *Condensing Intermittent Light, showing different Colours, with dark intervals.*—Equality of power can be produced by continuous straight prisms condensing the red arcs in the required ratio.

CHAPTER IV.

DISTINCTIVE CHARACTERISTICS OF LIGHTS.

AN increase in the number of lighthouses on any coast, leads necessarily to an increase in the number of distinctions. The following are the principal characteristics now in use, or which have been proposed, but there are some modifications and combinations of these, to which we shall also have occasion to refer.

I. The *Fixed Light*, which remains constantly in view, and invariably presents the same appearance, and is always of the same power in the same direction, whether it be condensed or naturally diverging in azimuth.

II. The *Double Light*, fixed or revolving, either shown from separate towers or from one tower at different heights.

III. The *Fixed Light, varied by Flashes*, which shows a steady fixed light, and at certain recurring periods becomes gradually for a few seconds of much greater volume and intensity.

IV. The *Revolving Light*, which at equal and comparatively long periods (such as once a minute), comes slowly and gradually into full power, and then as gradually disappears.

V. The *Flashing Light*, which at short periods (such as a few seconds) comes very quickly, though gradually, into view, and as quickly and gradually disappears.

VI. The *Coloured Light*, which is obtained by using

coloured media in connection with any of the varieties of characteristics described.

VII. The *Light of Changing Colour.*—See pp. 124, 131.

VIII. The *Intermittent Light,* which bursts instantaneously into full power, and after remaining as a fixed light of the same power for a certain length of time, such as one minute, is as suddenly eclipsed and succeeded by a dark period, such as a $\frac{1}{4}$ of a minute; and these phases, which occur at regular periods, are therefore unaccompanied by any waxing or waning, as in the revolving and flashing lights.

IX. *The Intermittent Light, of Unequal Periods,* which differs from the ordinary intermittent light just described, in showing from the same apparatus different durations of the light periods and different durations of darkness. Thus, for example, it may show a fixed light constantly for five seconds, then be eclipsed for two seconds; revealed again for two seconds, eclipsed again for two; and then continue steadily in sight for five seconds as at first.

X. The *Group Revolving and Flashing Light,* which differs from the common revolving and flashing lights by showing a group of waxing and waning flashes at unequal periods.

XI. *Separating Lights,* in which lights are made to separate and then gradually to coalesce either by horizontal or vertical movement.

I. *Fixed Lights.*—Nothing more need be said regarding the fixed light than has been already explained in previous Chapters, except to point out that, from its inferiority as regards power, it should not be adopted where revolving or condensing lights of greater power can be made available.

II. *Double Lights.*—The risk of confusion among fixed lights may be prevented by showing two lights, separated vertically from each other in the same tower, as proposed in 1810 for the Isle of May (Fig. 98) by Mr. R. Stevenson, who afterwards introduced this distinction at Girdleness, Aberdeenshire, in 1833. The effects of irradiation tend to blend together the visual images of such lights, long before their distance apart has become so small a fraction of the observer's distance from the tower as no longer to subtend the angle of the *minimum visible.* The distance at which the two lights appear separate can be found from the following table by Mr. Alan Stevenson,[1] founded on the fact that two lights 6 feet apart are seen just separate at a distance of a nautic mile.

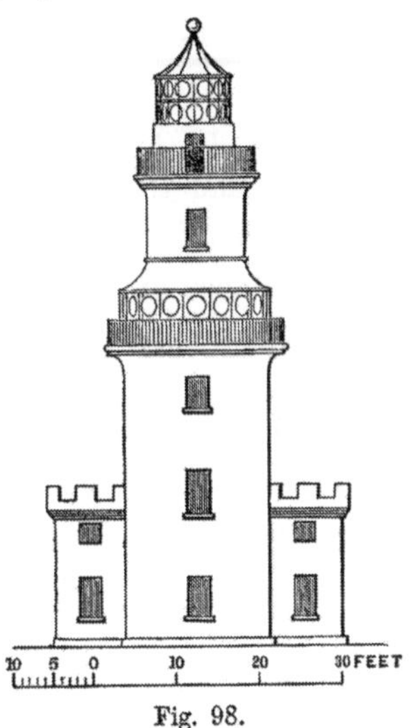

Fig. 98.

Distance of the Observer in Nautic Miles.	Vertical distance in feet between the Lights.	Distance of the Observer in Nautic Miles.	Vertical distance in feet between the Lights.	Distance of the Observer in Nautic Miles.	Vertical distance in feet between the Lights.	Distance of the Observer in Nautic Miles.	Vertical distance in feet between the Lights.
1	6·02	6	36·12	11	66·22	16	96·32
2	12·04	7	42·14	12	72·24	17	102·34
3	18·06	8	48·16	13	78·26	18	108·36
4	24·08	9	54·18	14	84·28	19	114·38
5	30·10	10	60·20	15	90·30	20	120·40

[1] Rudimentary Treatise on Lighthouses, Part III., 1850.

III. *Fixed Light, varied by Flashes.*—The mode of producing this distinction has already been described (Chap. II. **18**, **30**, 7). It is still largely used on the coasts of France, but has seldom been employed in this country. The defect, which is a serious one, is the great inequality of power between the fixed light and the revolving flash, for as the former can be seen for a much shorter distance, the distinction must cease at that distance, and the light then becomes simply a revolving one to an observer who is farther off.

IV. and V. *Revolving and Flashing Lights.*—The time during which an observer will continue to see a revolving or flashing light at each of its luminous periods will depend on the distance and the condition of the atmosphere. Hence some variation in the duration of the light-period is unavoidable in all lights which wax and wane, and which are therefore of unequal strength throughout their flashes. But in each case the *times of appearance* of the maximum power of the flashes are nevertheless of exactly equal recurrence, whatever be the state of the atmosphere or distance of the observer. The varying duration of the flashes possesses, moreover, one advantage, for it gives the sailor a rough idea of his proximity to the lighthouse, which is always a matter of importance. The mode of producing these distinctions has been explained in Chapter II. The revolving light was used at Marstrand, in Sweden, prior to 1783, but it is not known by whom. The *flashing light* was proposed by Mr. R. Stevenson, who introduced it at the Rhins of Islay, in Scotland, in 1825. It gives a flash every 5 seconds, and its phases are very striking and effective. At Ardrossan Harbour Messrs. Stevenson introduced, in 1870, a flashing light of only *one second* period, but it was found to be unsatisfactory, and was altered to its present period of one flash in every two seconds.

Other Modes of producing Revolving and Flashing Lights.—Besides the common arrangement explained in Chapter II., the following methods have been proposed:—

1. *Reciprocating Apparatus.*—Captain Smith proposed, with a view to economy of construction and maintenance, that when only half the circle is to be illuminated the apparatus should move forwards and backwards over 180°. The objection is the varying inequality of the *periods* in the different azimuths.

2. *Mr. J. T. Thomson's Reciprocating Apparatus.*—This possesses all the advantages, without any of the disadvantages, of Captain Smith's plan. Mr. Thomson describes his contrivance in the Trans. Roy. Scot. Soc. Arts, 1856, p. 308, in which, by means of rods and a pin at F (Figs. 99 and

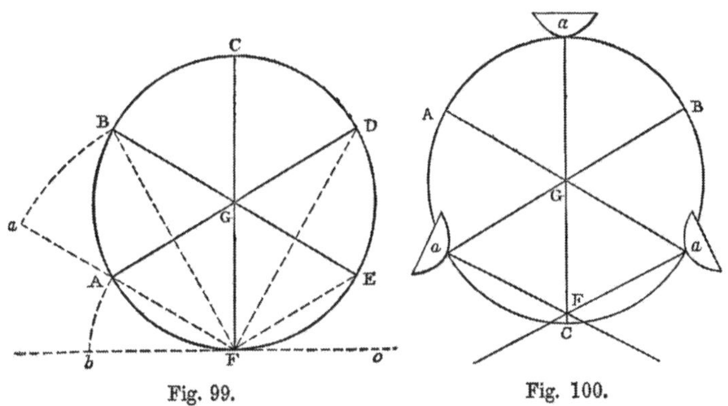

Fig. 99. Fig. 100.

100), the reflectors are reversed in position while traversing the dark arc.

3. *Reversing Apparatus.*—This plan, which I proposed in 1851, produces, in a different way, the same results as that of Mr. Thomson. In Fig. 1, Plate XXIV., *c* are the revolving holophotes, which have a vertical spindle passing through their centre of gravity, with a small pinion at their lower

end. E F are concave toothed segments, which turn each reflector over 180° as it comes round, so as to make them point seawards again, and in this way half the number of holophotes which would otherwise be required is sufficient.

4. *Fixed Lights illuminating the whole horizon, but showing Revolving or Intermittent Lights over small arcs.*—I have found that where only a small arc of a fixed light is to be made revolving or flashing, in order to cover an outlying danger, the effect can be satisfactorily produced by causing a straight upright mask to revolve horizontally on a vertical axis in front of an ordinary dioptric fixed apparatus. In this case the mask must subtend at the flame the same horizontal angle as that which is to have a characteristic different from that of the main apparatus. The dark periods will be extremely short at the limits of the danger-arc, and will gradually lengthen as the vessel nears the line of danger, which is in the centre of that arc. In Fig. 101

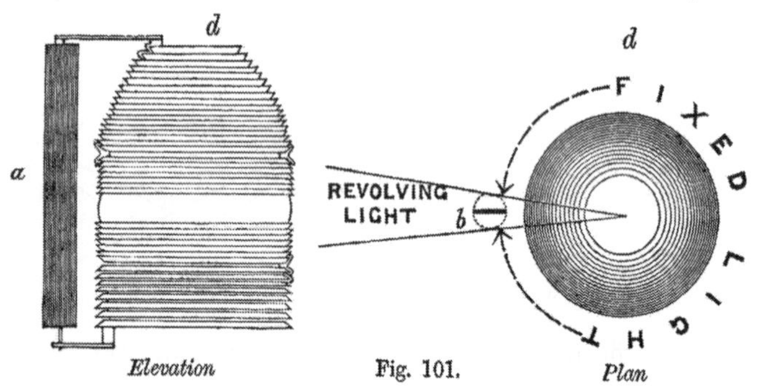

Fig. 101.

a is the mask in elevation, and *b* in plan, and *d* the fixed apparatus. Flashes of equal power may be produced, and thus made to show over the whole of the small arc, by using opaque leaves, *a*, Fig. 102, arranged either vertically or horizontally, like the Venetian blinds for house windows.

It is proposed to alter a sector of 79° at Dhu Heartach on this principle, in order to mark by red colour the extensive

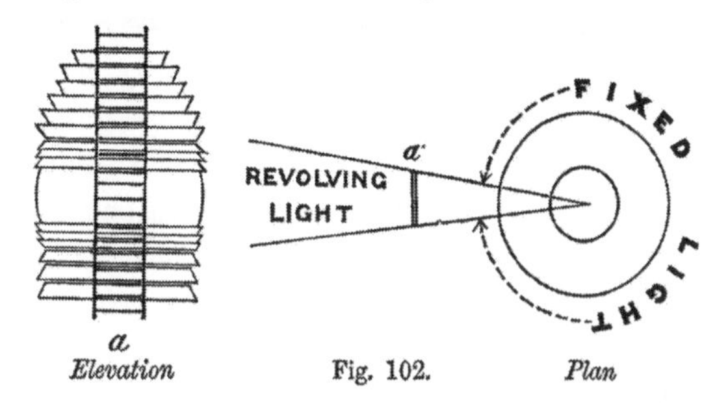

Elevation Fig. 102. *Plan*

and dangerous reef called the Torrin Rocks, which lie to the south of the Ross of Mull. The superiority of this mode of marking dangers, as compared with a fixed red sector, is about 4 to 1, for the full power of the light is preserved, while a red shade causes a loss of about three-fourths.

VI. and VII.—*Alternating Red and White Lights.*—The French alternating red and white light has an upper cupola of holophotal prisms, each of which parallelises 45° of light. The refractor and lower prisms give a fixed light. Each of the upper panel of prisms has a divergence of about 5°, and has a red shade placed in front. There also revolve, in front of the refractors and lower prisms, red shades, subtending an angle of 5°, being the same as the angle of divergence of the upper panels; and thus arcs of 5° of red and 40° of white are simultaneously illuminated, and shown alternately when the apparatus revolves, producing an alternation of red and white light of—say 40 seconds white and 5 seconds red, or any other periods in these proportions. But these proportions cannot be altered; and as the light in the axis of the red arc is stronger at the sides, the time during which the red will be visible must vary with the distance and state of the atmosphere.

Condensing Intermittent Light of changing colour.—The apparatus described at p. 124 are obviously very different from, and far more powerful than those last described, and are, besides, independent of the divergence of the flame.

VIII. *The Intermittent Light.*—The Catoptric intermittent light was introduced by Mr. R. Stevenson in 1830 at the Mull of Galloway. The occultations are effected by the sudden closing and opening of two intercepting opaque drums, Figs. 103 and 104, which inclose the apparatus,

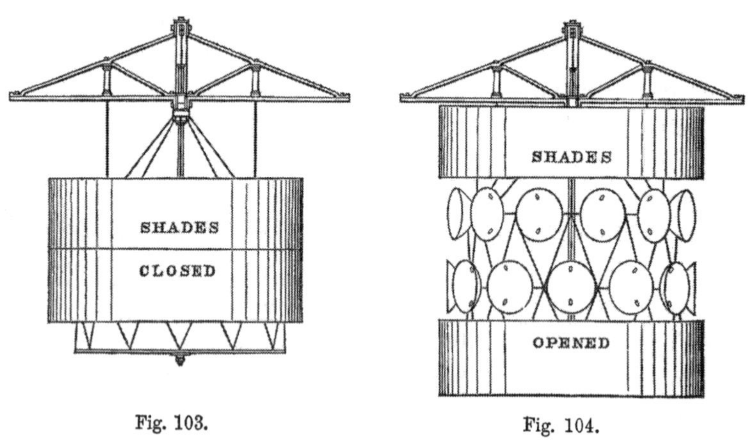

Fig. 103. Fig. 104.

and are moved vertically in opposite directions by means of machinery.

IX. *Intermittent Light of Unequal Periods.*

1. *Professor Babbage's Mode,* 1851.—The method proposed by Babbage for producing occultations was by successive descents and ascents of an opaque shade over the flame, instead of Mr. Stevenson's system of drums moving outside of the apparatus. The inequalities in the periods

were not introduced for producing optical distinctions, but in order to show the differences of numbers, as, for example, to indicate by a larger interval that three succeeding occultations meant 300, and not 30 or 3. A great practical objection to Babbage's proposal is the obstruction which it opposes to the access of the keeper to the burners for trimming the flames.

2. *Mr. R. L. Stevenson's Strengthened Intermittent Light*

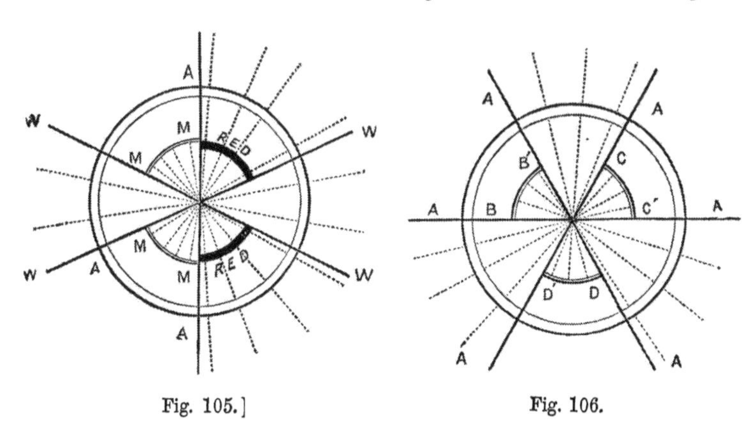

Fig. 105.] Fig. 106.

of Unequal Periods, 1871.—This was not only the first proposal for producing an optical distinction by unequal periods, but also the first proposal to save light in the dark sectors and to strengthen with it the light sectors, so that an existing fixed light could not only have its characteristic changed, but its power to a certain extent increased. This plan is produced by interposing

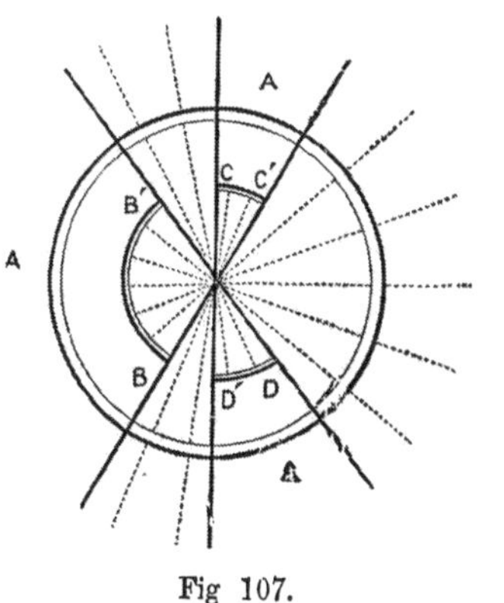

Fig 107.

between a fixed dioptric apparatus and its flame a spherical mirror, cut vertically or obliquely into separate sectors of such equal or unequal breadths as may be wanted, and thus coloured lights (Figs. 105), and intermittent or revolving lights of equal periods (Fig. 106), and of unequal periods (Fig. 107), may be all produced from the central flame, while advantage was for the first time taken of the spare light which spreads uselessly over a dark sector by strengthening a light sector in another direction, so far as the natural divergence of the flame admits. In this way a fixed light may have its power doubled—less the loss due to the spherical mirror. In Fig. 105, M M are the portions of spherical mirror, A W the strengthened red arcs, and W W the unstrengthened white arcs. In Fig. 106 C C′, B B′, and D D′ are the mirrors. In Fig. 107 B B′, C C′, and D′ D, are the mirrors which subtend the larger and smaller arcs.

3. *Dioptric Condensing Intermittent Lights*, 1872.—In the holophotal condensing intermittent apparatus (Chap. III. **14**) the power, as has already been explained, may be exalted by simply varying the duration of the light period to the dark, so that the power is in each case *inversely proportional* to the duration of the dark period.

4. *Sir Henry Pelly's Form of Intermittent Light.*—The occultations are produced in this apparatus by causing opaque shades to circulate *outside* of a stationary fixed dioptric apparatus. Sir William Thomson also proposed that their inner surfaces might be silvered, in order to save the light falling on them, by reflecting it back again through the fixed apparatus to the flame, so as finally to pass through and be parallelised by the other side of the apparatus. This plan is obviously not so good as the previous designs of Mr. R. L. Stevenson, in which the rays have to pass through

only one side of the fixed apparatus instead of through two, or as the condensing apparatus (Chap. III. **14**, 2) in regard to waste of light, while the power cannot, like it, be increased in any required ratio, but is dependent solely on the natural divergence of the flame. There is also the waste of light due to the spherical mirror in Mr. R. L. Stevenson's, and to the cylindrical one in Sir W. Thomson's, arrangement.

X. *Mr. Wigham's Group Flashing Gas Light.*—This useful distinction was originated by Mr. Wigham, in Ireland, in 1878. The effect is obtained not by optical combination, but by operating mechanically on the source of light, which is a gas flame placed in the centre of 8 revolving lenses. This mode of producing occultations by gas was first introduced in 1827 by the late Mr. Wilson at Troon harbour, but not in connection with revolving apparatus. At Galley Head the gas is continually turned up and down by clockwork while the lenses revolve, so that the ordinary flash is broken up into a series of shorter flashes. The *automatic gas meter* (Chapter V.), in which the flow of the gas itself regulates its supply, so as to produce an intermittent or revolving light, is however a simpler method of giving the alternations than any kind of clockwork, and could certainly be advantageously applied in connection with Mr. Wigham's plan. The objection to the mode of varying the flashes at Galley Head, which does not however appear to have been found of practical importance by the sailor, is, that the effect varies at the periods of raising and lowering the flame, depending on the position of the revolving lenses at the time. This was pointed out in Messrs. Stevenson's report of 9th November 1870, and also by Dr. Hopkinson, who remarks, "In the case of an observer at a little distance off, the flash might reach him when the gas had just been turned down, and then he would receive only 6 flashes," instead of 8 in the period of a whole revolution of the flame.

Dr. Hopkinson's Group Revolving and Flashing Lights. —This distinction also has been found to be well marked and effective. The apparatus is shown on Plate XXII., Fig. 3. Instead of the ordinary long flash produced by a large lens when revolving, the beam is split up by two or three portions of lenses, as may be desired, so as to give two or three flashes in rapid succession, after which there is a longer period of darkness. The means of producing this effect is optically simple, and it is also not very expensive, and has already been introduced successfully at five different places. An objection to this plan is the inequality of power of the flashes, when more than two are adopted, arising from the variation in the angles of incidence on the different lenses; but it does not appear to have been found a practical evil.

Mr. Brebner's form of Group Flashing Apparatus (Plate XXXV.)—In this better form of apparatus the lenses are cut horizontally in such proportions as, with the help of the upper prisms for the one half and the under prisms for the other half, to produce equal flashes. If then these two portions be separated horizontally, a small arc of darkness will intervene between the two flashes; or, as is shown on the diagram, the upper and lower prisms may be made to act together to produce one flash, and the lens, reduced in size so that its light shall be equal to the combined light of the two sets of prisms, to produce another. The axes of the two different sections should be inclined to each other at an angle of 10° to 12°.

XI. *Separating Lights* (Fig. 108).—In 1873 I suggested the employment of a wholly different element, viz. *the relative motion of two lights*, which is obtained by causing two fixed light apparatus, each illuminating the whole horizon, to revolve horizontally round a common centre. At distances sufficiently near, the appearance is that of *one* light separating gradually into two, which two again approach each other till they coalesce into one as before, a characteristic which is preserved in every azimuth.

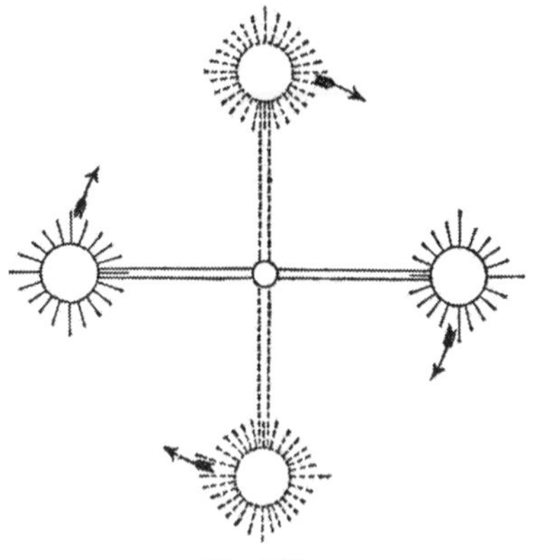

Fig. 108.

XII.—Distinctions by Colour.

1. The first proposal to adopt coloured media for producing distinctions in lighthouses seems to be due to Benjamin Mills of Bridlington, who introduced red shades at Flamborough Head, some time prior to the year 1809. The optical action of coloured shades is to prevent to a great extent the transmission of all the colours of the spectrum, excepting the one which is wanted. But in every kind of stained glass more rays than those of the one colour which is required are transmitted. This is proved by the fact that if white light be passed through two red panes, the emergent rays will be of a deeper and purer hue than after passing through a single pane. If the first pane had transmitted none but purely red rays, the inter-

position of a second red pane, though it would have lessened the intensity, could have had no effect in making the emergent light redder than before. But by adding a second pane, more of the other rays are screened out, and thus a purer red is obtained. From the greater loss of light by green and blue, red is the only colour employed, unless in certain cases where the range is very short, or where condensing apparatus is employed for increasing the power, when green is sometimes resorted to. Wherever white and coloured flashes are to be shown alternately to great distances, they cannot be seen equally far, unless they be derived from sources of illumination of different intensities. The first red and white light in which the transmitted beams were approximately equalised was the Bell Rock in 1842. But with such an apparatus, where the flash is produced by several independent reflectors, it is impossible, as already explained, from the optical nature of these instruments, exactly to equalise the two beams, and the usual size of the lightroom also precludes the employment of the additional number of independent lamps and reflectors which would be necessary in order to insure perfect equalisation; whereas, when there is only one central burner this difficulty disappears, for the apparatus can be divided as easily into unequal as equal sections, so that the red panels can be made to subtend a larger horizontal sector of the light than the white. It is however quite a possible case that in certain conditions of the atmosphere the equalisation of the red and white beams may be destroyed, and that one or other colour may prevail.

2. *M. Reynaud's Experiments on Coloured Lights.*—The most exhaustive investigation of this subject is due to M. Reynaud. The more important of his results, obtained with lights of equal initial power, are given in the following table:—

COLOURED LIGHTS OF EQUAL INITIAL INTENSITY.	Ratios of intensity at different distances.		
	From ·2 metres to 2 metres.	From ·2 metres to 6 metres.	From ·2 metres to 8 metres.
Red (of Gold)	1·46	2·12	2·34
„ (of Silver) ordinary	1·31	2·11	2·16
„ (of Copper) very dark colour	2·32	5·16	6·89
Green ordinary	0·85	0·68	...
„ deeper	0·77	0·61	...
Blue	0·80	...	...

This table shows that as the distance increases from which the light is viewed, the ratios of intensity of different red lights vary with the deepness of the colour, and that from copper, which absorbs most of the light at short distances, is most effective as the distances are increased. Other observations were made at greater distances than those in the table, and showed, 1st, At a distance of about 500 metres the white light ceased to be visible, while the reds excepting that of gold were still quite bright. 2d, At 750 metres the same lights remained visible, but the red was distinct only in the very deeply tinged glass produced by copper. All these experiments were made when the atmosphere was clear, but similar observations were also made during a fog with the following results:—Of five reflectors, all of equal power (about 60 carcel burners), the first was uncoloured, the second was a red produced by gold, the third a red by copper, the fourth green, and the last blue. The white and red lights became invisible at a distance of 1600 metres. The red of gold was with difficulty distinguishable at 1500 metres, while that of copper was still well defined. The green disappeared at 1000 metres, and the blue at 530 metres. In other words, if red and white lights appear of equal power at short distances, the red will be seen at a greater distance in a fog. The reverse is true of the green and blue, which diminish in intensity far more rapidly than the white as the distances are increased, and still more so in a thick atmosphere. In the French lights they use red glass tinged with copper, of such a shade as not to reduce the light more than from $\frac{1}{2}$ to $\frac{2}{3}$ds. The amount of white light which is in practice allocated to the red to produce equality is from two to three times that of the white. But the results of the experiments would seem to justify a smaller disparity than this.

3. *Messrs. Stevenson's Experiment on the Power of Red and White Lights.*—The apparatus used in these experiments for comparing the power of the lights was suggested by Mr. Alan Brebner, and consisted of two first-order lenses, with a four-wick colza burner in the focus of each. The corners of the lenses were covered up, so as to leave circular discs of light, whose diameters were equal to the breadth of the lenses. Over the disc of the white light lens, there was placed a fan, *a b c* (shown in Plate XXII., Fig. 1), made of thin sheet iron. The centre of the fan was coincident with the centre of the lens, and it was capable of being spread out so as to cover up the whole disc of white light.

The arms of the fan were sectors of the circle, so that each cut off from the whole, a portion which contained weak and strong light in the same proportion as the whole disc. It must also be understood that the burner did not come in the way of any part of the lens, which was therefore illuminated equally over its whole surface.

The light of one of the lamps was made red (Plate XXII., Fig. 2) by one of the ordinary red glass chimneys.

The lights were exhibited from lanterns at Granton, near Edinburgh, the distance betwixt them being about 100 feet, and the observations were made from the Calton Hill, $2\frac{1}{2}$ miles distant, at a point to which the two lenses were directed.

When the two lights were first exhibited in their normal condition, that from the white lens appeared greatly larger and more powerful. The fan was then spread out over the white lens and the effect observed, the result being that the white maintained its superiority until $\frac{3}{4}$ths of the circle was covered up, when the two lights appeared about equal, thus showing 4 to 1 in favour of the white.

4. *Experiments on the Penetrative Power of a White and Red Light.*—A light from a gas Argand burner was placed at each end of the scale of a bunsen photometer; one of these lights was made red by means of a red glass chimney, and the other had the usual white glass chimney. In this state the two lights were equalised, so that the disc of the photometer stood exactly midway between. The distance betwixt the flames was 90 inches. A small trough, made of plate glass, was next filled with a mixture of milk and water, so as to endeavour to resemble a fog, and when this was placed next the red light the white light was 2·015 times stronger, and when it was placed next the white light the red appeared 1·96 times stronger. With another mixture containing a larger proportion of milk placed next the red, the white was 5·22 times stronger, and when placed next the white the red was 5·006 times stronger. After this the red glass was taken off, and a white one substituted, when it was found that the light from which the red chimney had been removed was now 4·9 times stronger than the other. This last result

agrees pretty well with that obtained by the experiment with the fan, which was 4·0.

It thus appears from these experiments that the red glass which is used in the Scotch service absorbs about $\frac{3}{4}$ths of the rays, or, in other words, the red must have an initial intensity *four times greater than the white* in order to produce equality. If it were warrantable to regard the mixture of milk and water as possessing the same absorptive power as fog, it would further follow from the experiments that the penetrative powers of red and white were equal. But Reynaud's experiments give a different result, and if his be accepted as more correct, because made in actual fog, it follows that the milk-and-water mixture is essentially different from fog in its selective absorption, and the last experiments cannot therefore be founded on.

In order to equalise red and white lights I am inclined to think that the relative amounts of light required for that purpose should either be spread equally over equal areas, or else over areas of corresponding intensities, for it is quite possible that the ratio of loss in passing through the atmosphere may vary with the intensity of the light, and be less for an intense than for a more diluted beam of rays.

The results which have been given are certainly of a very conflicting nature. At the Bell Rock the red and white lights have been so nearly equalised that, as seen from Arbroath, 11 miles distant, from June 1876 to January 1880, being in all 1096 observations, there was only *one* occasion, at the times when observations were made, on which the white was seen when the red was not. The ratio of red to white at the Bell Rock is 1·7 to 1. The Wolf is in the ratio of 2·233 to 1. The French in the ratio of 3 to 1, while the experiments with the fan gave the ratio of 4 to 1. Making every allowance for the differences that may exist in the tints of the different glass shades that were tried, there seems to be great need for a further set of experiments on this important subject.

XIII.—Mode of Producing Coloured Lights.

1. Where the back rays are not acted on by a spherical mirror and made again to enter the flame, but are passed at once to the apparatus, and thence to the sea, the best method of colouring the light is to tinge the glass chimney with the required colour. But many coloured lights require some of the rays to be returned by a spherical mirror through the flame before they pass out seawards. In such a case the emergent light will not be all of one shade, for the portion that passed *twice* through the coloured chimney will have attained a deeper shade than that which passed only once through one of its sides. So that if those rays which pass once through the chimney be sufficiently distinctive, which they ought to be, then it follows that there is an unnecessary loss of light occasioned by the transmission of any of the rays a second time through the coloured chimney. In order to avoid this loss it is better to use a colourless chimney, and to place plain coloured shades either immediately in front of the apparatus itself or else close to the panes of the lantern of the lightroom. In the latter case, owing to the close proximity of the shades to the lantern, there is often imperfect ventilation, and in certain states of the weather atmospheric condensation takes place, involving a very great loss of light.

If, as has long been practised in the preparation of certain optical instruments, two pieces of glass be united together by means of a transparent cement which has nearly the same index of refraction as glass, a ray will pass through such a composite substance without suffering any notable refraction at the inner surfaces; and on looking through such a combination of materials the two inner surfaces of the plates of glass will be invisible, simply because the light passes from the first piece of glass into the cement, and from the cement into the second piece of glass, without suffering any appreciable change in its direction, any more than it would had it passed through a solid piece of homogeneous glass of the same thickness as the compound mass. Such a cement is, as already stated, the well-known Canada balsam, which has very nearly the same index of refraction as crown glass.

If plates of coloured glass be cemented to the outer faces of lighthouse apparatus the rays will therefore pass from the glass of the apparatus through the Canada balsam into the coloured glass, practically speaking, without refraction, and, therefore, without appreciable loss by superficial reflection, and in this way the loss at two surfaces will be saved. Taking the loss for a polished surface at $\frac{1}{30}$th of the whole, probably $\frac{2}{30}$ths or $\frac{1}{15}$th of the incident light would be saved. But this is not the whole gain, for the kind of glass which is employed for colouring cannot by the present mode of manufacture receive a polished surface on both its sides, so that the real loss due to irregular reflection and scattering must be considerably more than $\frac{1}{15}$th. Count Rumford[1] found that the loss in passing through "a very thin pane of clear white or colourless window glass, not ground," varied from ·1213 to ·1324; so that, allowing for the loss by absorption in passing through this "very thin"

[1] Philosophical Papers by Count Rumford. London, 1802. Vol. i. p. 299.

pane, we shall certainly save more than $\frac{1}{10}$th of the whole incident light. In order, however, to take proper advantage of this mode of combination, it is necessary that the coloured side of the glass should be placed next to the outer face of the apparatus, so as to admit of the outer or uncoloured surface of the shade being polished—a process which can easily be accomplished after the plate has been fixed to the prism by the Canada balsam. It may perhaps be supposed that there will be an increase in the amount of polishing, because the outer surface of the coloured glass will require to be polished as well as the surfaces of the prisms themselves. This, however, is not necessary, because, although the outer face of the prism be left unpolished, the Canada balsam which cements the pane of thin coloured glass will, by its practical abolition of the surfaces, render the irregularities of no real consequence.

It has been found that this mode of producing coloured light is efficient for surfaces of moderate size, but when a large pane was tried, separation was found to take place after a lapse of a year or two. Messrs. Barbier and Fenestre of Paris have constructed, apparently with considerable success, pieces of white and red glass united together by fusion.

2. *Coloured Silver-plated Reflector.*—Another mode of producing coloured light is by means of reflectors constructed of small facets of mirror glass. As already noticed, these facets may be easily made of tinted instead of white glass, so that no coloured chimney is required.

XIV.—Colour-Blindness.

An objection to all coloured lights is founded on the fact that there are individuals who are unable to detect any difference between some one colour and another. The late Dr. George Wilson, of Edinburgh, was one of the first writers who went fully into the statistics of this physical imperfection. Professor Holmgren of Upsala has recently subjected from 60,000 to 70,000 persons in Europe and America to the test of colour-perception, by requiring them to match differently tinted wools. The results of these inquiries, which confirm that of the earlier writers, is, that on an average 4·2 per cent (ranging from 2 or 3 to 5 per cent of males) were congenitally colour-blind to a greater or less extent. It unfortunately happens that the defects from normal vision (or the power of discerning the difference between all colours) is generally an inability to distinguish between red and green, which, as we have said, are the only tints that are employed for lighthouse distinctions. There can be no question that this physical defect is an objection to employing colour distinctions, yet most sailors are not so disqualified, and to them will generally be intrusted the important duty of looking out for the different sea marks when the vessel is seeking the coast.[1]

XV.—*Professor Babbage's numerical system of distinguishing Lights, as modified by Sir William Thomson* 1873.—The essential principle of the simple characteristics at present in use is that of optical distinction, and of strongly marked, and therefore very obvious divisions of time, in the exhibition of the different phases of light and darkness; while that advocated by Sir W. Thomson is the alteration of all fixed lights

[1] The Board of Trade, in 1877, introduced the "colours test" in the examinations for masters' and mates' certificates, and in two years it has been found that about half per cent (·43) of those examined were decidedly colour-blind. The Board of Trade has very judiciously come to the important decision that all masters and mates must now pass the "colours test" as a preliminary part of their examinations for certificates in navigation.

to rapid eclipses, which, it is understood, are intended to represent letters of the alphabet, as in the Morse system of signalling. Sir William Thomson thus describes his proposal:—"To distinguish every fixed light by a rapid group of two or three dot-dash eclipses, the shorter or dot, of about half a second duration, and the dash, three times as long as the dot, with intervals of light of about half a second between the eclipses of the group, and of five or six seconds between the groups, so that in no case should the period be more than ten or twelve seconds."

At present the sailor, when falling in with a light, at once recognises it as fixed or revolving, or revolving red and white; in periods, such as once a minute, or once in half a minute; or flashing, in such short periods as from five to twelve seconds; or as intermittent, showing for a certain length of time a steady fixed light, which is instantly succeeded by a dark period, and therefore readily distinguishable from the revolving, by the much longer duration of light, and by the suddenness of the occultations. Or, again, the intermittent condensing light can be made (Chap. III. **14**, 2) to exhibit certain intervals of fixed light of unequal duration; and lastly, the group flashing revolving lights exhibit two or more flashes, appearing at short intervals from each other, which are succeeded by longer intervals of darkness.

In order to recognise lights such as these, any appeal to a watch is hardly required, because the characteristics are either purely optical, or the periods are so widely different as to be at once apparent.

The origin of proposals to alter the present system rests upon an entirely erroneous conception regarding facts which are well established. The discussions on the subject are leading to a widely diffused notion that the great cause of shipwrecks is the mistaking one light for another by the mariner. Mr. Alan Stevenson, in an exhaustive report in 1851 on Babbage's proposal, showed by actual statistics that for the four years immediately preceding that report the true cause of shipwrecks at night, on the coasts of Scotland, was the non-visibility of the lights, and not the confounding of one with another. Out of 203 shipwrecks, 133 occurred by night; and in *only two of these* was it even *alleged* that the lights when seen were not recognised. Mr. Stevenson showed that the sailors must in both cases have been grossly ignorant. In only one of these two cases were the lights specified that were said to have been confounded, viz. the *revolving* light of Inchkeith with the *fixed* light of Isle of May. If allegations of mistaking such radically different characteristics are to be counted of any value, what, it may be asked, would become of the Morse dot and dash system, with its intricate and minute distinctions?

It seems unnecessary to go over all the questions discussed in Mr. Stevenson's report, in which he clearly pointed out that the grand requisite of all sea lights is *penetrative power*, and not an increased number of different periods of the recurring appearances. He also shows, as a corollary, the fatal mistake of altering all the lights to the intermittent character, as proposed by Mr. Babbage, by which their power would be prodigiously reduced (Chapter II. 2). It must not however be supposed that the present distinctions are to be regarded as incapable of improvement or extension. I entirely agree with Sir W. Thomson in thinking it desirable to compress the whole features of distinction within the smallest space of time that is really consistent with clearness of recognition and simplicity of nomenclature. This had indeed to a large extent been carried out by the

Scotch Board after 1870, on the recommendation of their engineers,—that the slowly revolving and intermittent lights which were erected early in the century should be accelerated, in consequence of the much higher speed which has of late years been attained by ships, and especially by steamers. The Trinity House of London have also, it is understood, adopted the same improvements in the speed of revolution of the apparatus. I further concur, so far, in opinion with Sir William Thomson, as to think that fixed lights might in some cases be advantageously dispensed with. The best mode of altering the present fixed lights, without cancelling the dioptric apparatus, is to make them condensing-intermittent by means of revolving straight condensing prisms placed outside, which will not only alter the characteristic, but also, as has been shown, increase very greatly the power of the light. Though Mr. R. L. Stevenson's apparatus was proposed in 1871, and the condensing principle in 1872, Sir W. Thomson did not adopt either of these plans, but used Sir H. Pelly's plan of revolving opaque masks. The rays intercepted by such masks are therefore lost, and the light must remain of the same power as that of an ordinary fixed one, which has been shown to be the weakest of all. Even though the masks were made, as he once proposed, to reflect the rays back through the flame, the loss would still be great. His Belfast apparatus would certainly have been optically far superior, and therefore far more powerful, had the condensing principle been adopted. As regards the periods that may be employed for intermittent lights, it may be mentioned that, in the case of Ardrossan, already referred to in this Chapter, where a flash once in every second was tried, and the physiological effect found unsatisfactory, the cause of imperfection is sufficiently manifest; for it is obvious that the waxing and waning of the flash, due to the raising and lowering of the gas-flame, would reduce the duration of the light period. Although the time of recurrence of the flashes remained once a second, the *full power* of the light would continue for only a fraction of a second, and that fraction would vary with the state of the atmosphere and the distance of the observer. But this objection does not apply to lights in which there is no diminution of the light period by waxing and waning, for, if a one-second intermittent light be visible at all, it must be so uninterruptedly during the whole second. From actual trial with a condensing intermittent light, I am disposed to expect that a light-period, even of only a single second of duration, may be found practically available.

It is very difficult to discover on what principle the Morse alphabet is advocated as superior to all other modes of distinction. The preference of such a system cannot, as we have seen, be grounded on the facts supplied by shipwreck statistics. Reference has been made by the advocate of the system to the peculiar effect of a street lamp in London, which, owing to the presence of water in its supply pipe, produced a twinkling, intermittent effect, that arrested the attention of a passer-by much more than the monotonous effect of the others which were steady. But, as Mr. Morris has well observed, if *all* the adjoining lamps had twinkled, in slightly differing times, the attempt to distinguish one from another would surely have been very perplexing, if not absolutely bewildering. This case illustrates very clearly the real objection to the Morse alphabet system, which is not the exhibition *here and there* on the coast, as at Belfast and the Clyde, of a rapidly intermitting apparatus, whose phases are either equal or unequal, and which phases or the order of their succession need not be particularly noted by the sailor, but the *general* adoption of such lights, which would require a strictly accurate discrimination of all the phases exhibited at each station on the coast. The sailor, in coming over sea, knows, approximately at least, where he should be, and when he sees a flashing light, he at once remembers what light is flashing on that part of the coast, but if the Morse system were in general use, he would, without a book of reference and careful watching of the number and order of recurrence of the long dashes and short dots, be entirely misled, so as to mistake the light he saw for the one immediately adjoining, or else for some other not very far off.

The conclusion, therefore, which seems warrantable from what has been adduced is, that lights should be distinguished, as at present, either by purely optical characteristics, *i.e.* by appearances which are at once appreciable by the eye, or else by widely different periods of time, and not by minute optical phenomena exhibited in rapid succession, which require to be counted, or at least to be very carefully noted, so that their meaning may be understood. A certain intermixture of lights, showing minute rapid alternations of irregular periods, such as suggested by Mr. R. L. Stevenson, is however in no way objectionable, but, on the contrary, advantageous, so long as they are separated from each other by several other lights of the ordinary character, so that a careful study of their phases is not required. But rapidly intermitting lights closely adjoining each other, whether in narrow seaways or on the ocean seaboard, could not but occasion great anxiety, and probably lead to serious mistakes. I am therefore decidedly of opinion that our coasts should continue to be marked in the manner which is now in universal use throughout the world. Similar views have been very ably advocated both by the Trinity House and the Irish Commissioners.

CHAPTER V.

BEACONS AND BUOYS.

1. Our shores are marked for navigation in two distinct ways, one of which—the lighthouse—is of use by day and night; and the other—the beacon or buoy—by day only. Wherever a rock or projecting spit is too small to allow, or too unimportant to warrant the erection of a lighthouse with tower, apparatus, and lodging for the keepers, the engineer can only substitute a beacon or perch, consisting of a bare framework or skeleton of iron, which is usually surmounted by a cage for the use of shipwrecked seamen, should they be fortunate enough ever to reach it. This, however, is a very indifferent expedient, and does not satisfactorily indicate a danger except in daylight. But when the rock lies constantly under water, the means that can be adopted are yet more unsatisfactory; for, unless recourse be had to a floating lightship, equipped and maintained at great expense, a floating buoy, which is generally still less obvious than a beacon, was till lately the only alternative. The importance of exalting this subordinate class of indications to the rank of illuminated night marks, must be apparent to all who are acquainted with coast navigation. The great expense of erecting lighthouses like the Eddystone and Bell Rock, accounts for there being so few of such structures in existence. Hence many dangerous rocks are still unlighted,

and there are many more that are without even beacons or perches. It seems only natural that some share of the attention which has been so long and so successfully devoted to lighthouse optics should be bestowed on the marking of those dangers, and we will now explain the different methods which have as yet been either employed or suggested for increasing their efficiency.

Dipping Light.

2. In all the apparatus employed for lighthouses, the object, as has been seen, is to gather together the diverging rays, and to direct them towards the sea, so that the vertical axis of the apparatus and lamp is placed nearly at right angles to the plane of the horizon. The dipping light,. on the contrary, should have its axis inclined downwards at a certain angle, so that the rays, instead of being projected nearly horizontally, will be thrown downwards upon the sea as represented pictorially in Fig. 109, and in Figs. 110

Fig. 109.

and 111, which show a dipping light in plan and section. A represents a lighthouse placed upon a headland ; B a sunken rock; C C, C C, the outline of the nearest safe *offing*, within which no vessel should come during the night. If the lamp

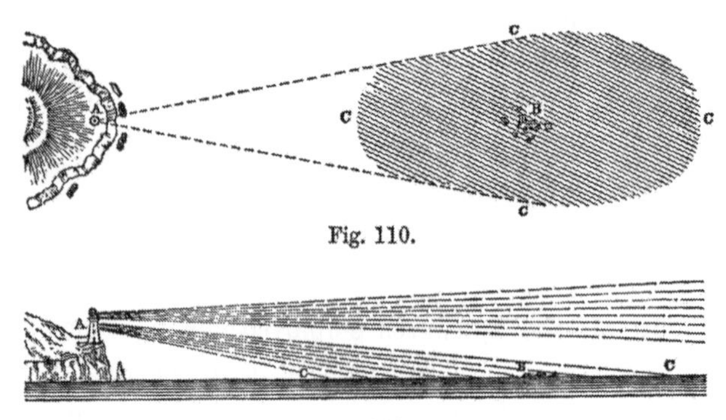

Fig. 110.

Fig. 111.

and reflector are inclined at such an angle to the horizon as shall throw down the rays on the space C C, C C, then whenever a vessel makes this light it is time to *put about*, as the margin of safety has been crossed. When dioptric apparatus is employed, each part must therefore be made to give the rays the required dip in the vertical plane.

The dipping light is on the same principle as a night-signal which was proposed by Mr. Alan Stevenson in 1842 for use in railways,[1] by which the engineman was warned of his approach to the station whenever he crossed the dipping beam of rays. It is said that a light upon this principle was many years ago used at Beachy Head, but it was not found to answer the purpose. Very probably the difficulty arose from too great divergence, that fertile source of trouble in all apparatus. If, for example, a small-sized holophote has its axial beam pointed somewhat seawards of the reef, so as to throw the strongest light in that direction, the base of the diverging cone will then cover many miles

[1] Trans. Roy. Scot. Soc. of Arts, vol. ii. p. 47.

of sea beyond the safe offing, unless the apparatus has a great dip, which is only possible when the tower is very high, or where the danger is close to the shore. No doubt the small divergence of the electric light would greatly lessen the evil; but there are not many situations where such a mode of illumination is likely soon to be provided.

The following modes of reducing the divergence of the light may be found useful where the land is low and the danger lies far off the shore. If a holophote (Plate XXXII. Fig. 3) be inclined to the horizon at a certain angle, and a series of reflectors or prisms be placed above the ground on a stage, the parallel rays from the holophote may in many situations be sent down on the sea by the reflectors at the required angle. The amount of divergence can also be exactly adjusted in amount by increasing or diminishing the distance, because all rays which have greater divergence than is wanted will escape upwards, while those only which have the proper amount will be thrown downwards on the distant danger. The same result may also be produced by pointing the axis of the holophote directly to the place of danger, in which case the amount

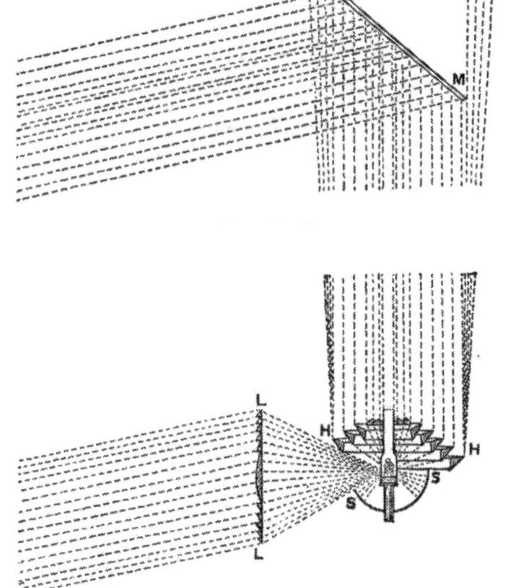

Fig. 112.

of divergence can be obtained by interposing between the apparatus and the sea at the proper distance a mask or screen, having an aperture made in it through which those rays will pass which have the required divergence, and all the others which have more divergence will be stopped by the screen. Another device is represented in Fig. 112, in which L is a lens with its axis pointed to the danger, H H, S S, a holophote pointing upwards, and M M, a plain reflector throwing only the slightly divergent rays downwards on the sea. In order to prevent confusion in the diagram, those rays only which diverge from the edge of the holophote are shown.

Apparent Light.

3. There are many places where there is little sea-room, and the danger, whatever it be, must be passed by vessels at a short distance. In this class of localities are included rocks at the mouth of bays and roadsteads where the fairway is not broad, narrow Sounds, and the entrances between the piers of harbours. In cases such as the last named, sometimes one pier-head must be hugged, sometimes the other, according to the direction of the wind; and so critical in stormy weather is the *taking* of a harbour that even a single yard of distance may be of consequence. Those only who know from actual experience the anxiety which is felt in entering a narrow-mouthed harbour under night, with a heavy sea running, can appreciate the vital importance of knowing at as great a distance as possible the *exact* position of the weather pier-head.

The dazzling reflections of the sun's rays, which are often produced by fragments of glass or glazed earthenware when turned up by the plough, are accidental examples of the mode of illuminating beacons placed on rocks at sea, which I proposed in 1850;[1] for the *apparent* light consists of certain forms of apparatus for reflecting and redispersing at the beacon, parallel rays which proceed from a lamp placed on the distant shore. By thus redispersing the rays an optical deception is produced, leading the sailor to suppose that a lamp is burning on the beacon itself. The first light of this kind, shown pictorially in Fig. 113, was established

Fig. 113.

in 1851, in Stornoway Bay, a well-known anchorage in the Island of Lewis, possessing shelter which has long been highly prized by the shipping frequenting those seas. It measures about a mile in length, and its entrance is about half a mile in breadth; but available sea-room for a vessel entering is materially reduced by a submerged reef, B (Fig. 114), lying off Arnish Point, on the south side of the entrance. For the purpose of rendering this sheltered anchorage safely available alike by night as by day, a re-

[1] Trans. Roy. Scot. Soc. Arts, vol. iv.

volving light was erected on the Arnish Point, A, and a beacon and apparent-light apparatus were placed on the sunk reef which lies off the Point.

The erection of a lighthouse for opening up this valuable harbour of refuge was long talked of; and the danger of the

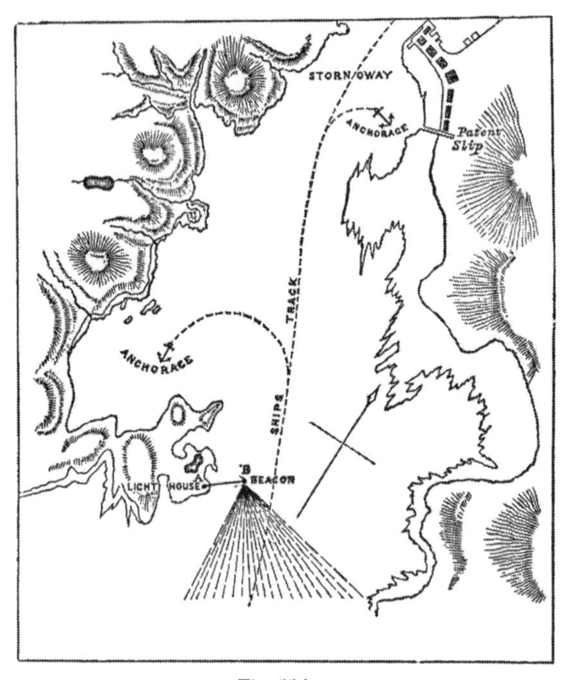

Fig. 114.

Arnish Rock was so much felt that it was proposed that the tower, instead of being erected on the mainland, should be built on the rock itself, which would of course have been attended with very great expense, for the reef is exposed to a heavy sea, and is quite inaccessible, unless by a boat or at low water of spring tides. The diverging lines represent the directions of the rays after being reflected at the beacon, while the thick line A B is the beam of parallel rays proceeding from the lighthouse on the shore. The revolving apparatus for the general use of seamen navigating the coast is placed in the light-room at the top of the tower on the shore, and the apparatus for throwing the light on the beacon at sea is placed in a window at the bottom. The beacon is 530 feet distant from the lighthouse, and is a truncated cone of cast iron, 25 feet high. On the top of the beacon a small lantern is fixed, containing the optical part of the apparent light, the centre of which is exactly on the level of the axis of the apparatus on the shore.

In the Notes which were appended to "The Holophotal System of Illumination," in the Trans. Royal Scottish Society of Arts for 1850, the possibility of employing apparent lights, and different methods of constructing them, were proposed. It will, however, be enough to describe the mode which was ultimately found best at Stornoway, where, for nearly 30 years, the apparent light has given perfect satisfaction to seamen. The optical arrangement on the beacon,

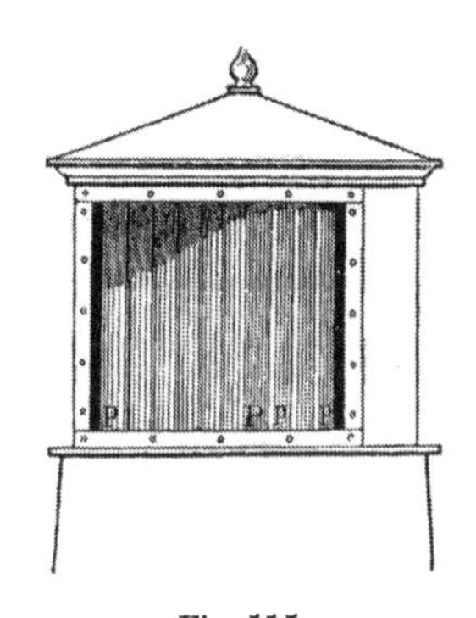

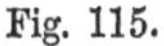

Fig. 115.

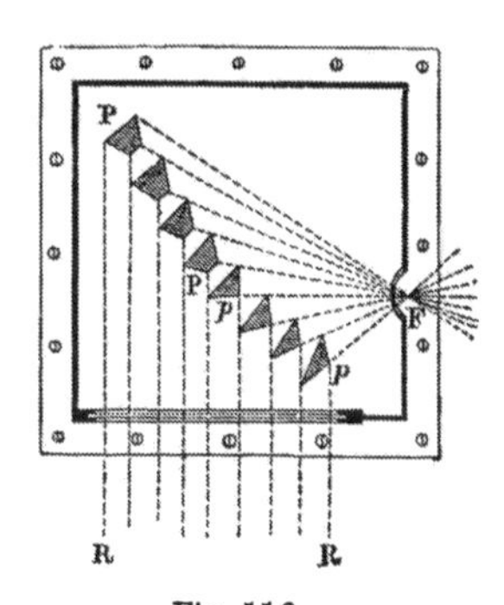

Fig. 116.

shown in Figs. 115 and 116, consists of straight "back" prisms P P (Chap. II. **30,** 9), and straight prisms of the ordinary form *p p* (Figs. 115, 116), both of which cause the beam to diverge by single agency over the seaward arc of 62°.

Apparent lights have since been established at Grangemouth 535 feet from the light, and at Ayr 245 feet from the light; and also by Messrs. Chance, for the Russian Government at Odessa, in the Black Sea, which is 300 feet from the light, and at Gatcombe Head, Queensland, where the intervening distance is also 300 feet.

The peculiar advantage which the apparent light possesses over both the electric light by wires and the gas light by submarine pipes, both of which will be afterwards described, is its absolute freedom from risk of extinction; for the means of illumination being on the shore are always accessible. Unless the beacon itself is knocked down by the waves, no storm or accident can interrupt the exhibition of the light.

Designs for increasing the Density of the Beam of Light sent to the Beacon.

4. The great difficulty to be surmounted in the apparent, and to a less extent in all other lights, is the diffusion of the rays by divergence, for there is only a very small proportion of the light that really falls on the beacon. The electric is clearly, therefore, of all others, the best luminary for the purpose, but the cost of maintenance has hitherto prevented its adoption for apparent lights.

One mode of increasing the intensity of the light on the shore is to increase the density of the luminous beams, by causing different flames to converge to the same focus. Brewster (Edin. Trans., vol. xxiv., 1866) gave two designs for this purpose, though not with reference to their application to apparent lights. In his catoptric design a flame was placed in one of the foci of an ellipsoidal mirror, by which the light was conveyed so as to strengthen another flame in the conjugate focus, which was also in the focus of another apparatus in front of it, but all the light which escaped past the lips of the reflector was lost, while in his dioptric design all the light was lost except the small cones which fall upon the lenses, as shown in Fig. 117, in which R S is a spherical mirror, A B, B C, etc., are the lost rays, and R R show the balance of the light which emerges in a beam of parallel rays.

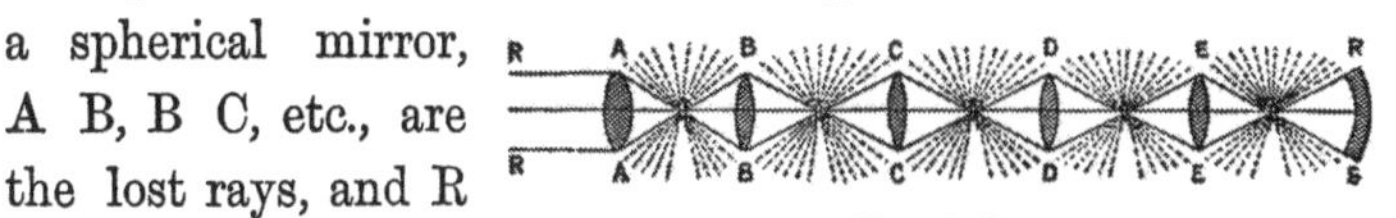

Fig. 117.

Such arrangements as these were therefore very far from being holophotal, and Sir David states indeed that "*the holophote principle is inapplicable*" to such a purpose. It is quite true that, besides the waste of light from employing agents that are not needed, the mode which he adopted for parallelising the rays could not possibly effect the purpose. But if the correct holophotal principle, already explained (Chap. II. **30**) be employed, and the unnecessary agents abolished, the problem is easily enough solved, and the light from any number of flames can, geometrically at least, be conveyed to the primary flame. In the paraboloidal holophote, A, Fig. 118, all the rays proceeding from the primary flame are directly parallelised, and sent forward upon the distant beacon (1*st*) by the lens L, whose principal focus is in the flame; (2*d*) by the paraboloidal strips P; and (3*d*) by the spherical mirrors, with the exception of the small cone P′ F P′, which is not intercepted, but is allowed to escape backwards through a hole in the mirror, and is finally sent back again to F, after being acted upon by the ellipsoidal holophote B, in which F′ is an auxiliary flame,

L′ a lens having its conjugate foci in F′ and F, ellipsoidal

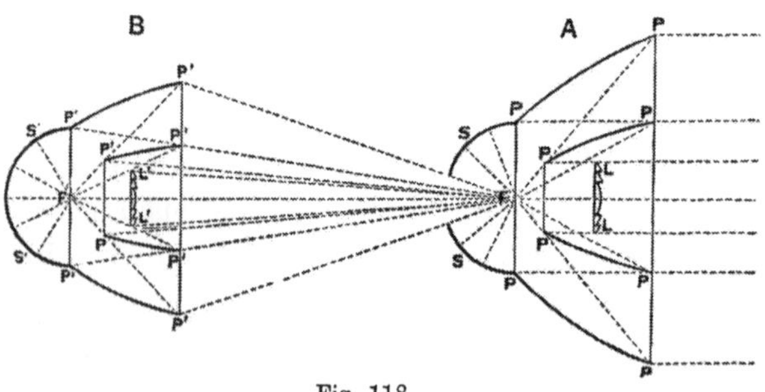

Fig. 118.

strips P′ having their foci in F′ and F, and lastly a spherical mirror S′. All the rays proceeding from the auxiliary flame F′ are therefore conveyed in a dense cone to F, which is at once the centre of the mirror S, the conjugate focus of the lens L′, the principal focus of the lens L, one of the foci common to the ellipsoids P′, and lastly the common focus of the paraboloids P. This apparatus, which may be constructed entirely of glass, can be applied to any number of auxiliary flames placed in line behind each other, all the rays from which will be ultimately parallelised in union with those from the original flame F. It was found, in constructing a catadioptric apparatus of the kind (Fig. 118), that, from the imperfections in the form of the reflectors (which were of a somewhat temporary character, and made of very light plated copper), combined with the small focal distance compared with the diameter of the oil flame which was employed, the rays were not so accurately converged as to justify its adoption. With more carefully constructed optical agents, or with the electric light as the illuminant, the result would unquestionably be very different.

Another plan consists of a single ellipsoidal mirror truncated at both ends, with a holophote and a spherical mirror attached, and a small lens having its conjugate foci coincident with the foci of the ellipse, as shown in Fig. 119.

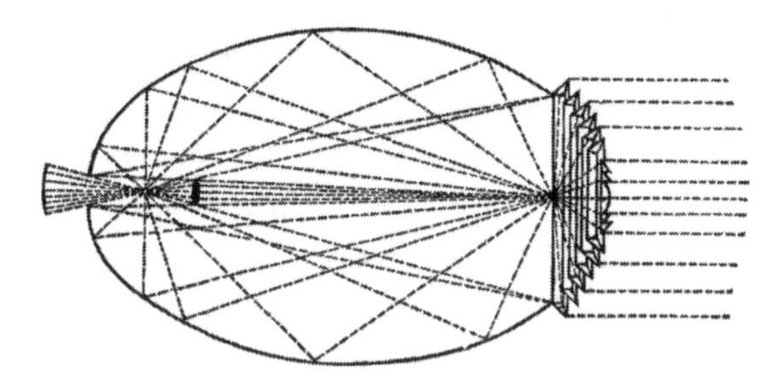

Fig. 119.

The action of this instrument will, after what has been described, be obvious on inspection.

Illumination by Electricity conveyed through Submarine Wires.

5. Before adopting the principle of the apparent light for Stornoway, it occurred to me (Trans. of the Royal Scottish

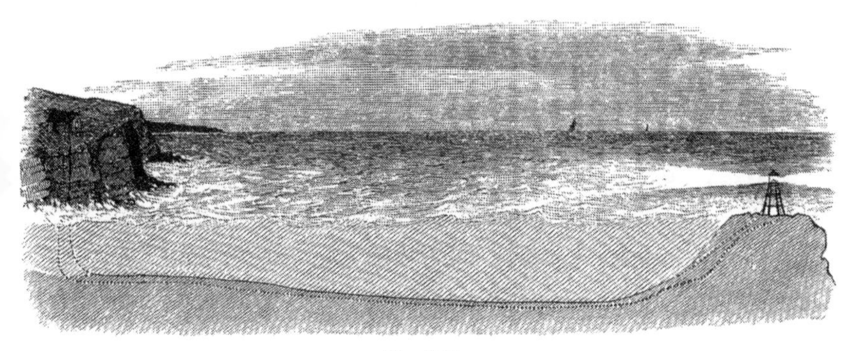

Fig. 120.

Society of Arts, vol. iv., 1854) that it might be possible to illuminate beacons and buoys by electricity, conveyed through

submarine wires, as shown pictorially in Fig. 120, but I dismissed the plan as inapplicable in the then existing state of electrical science.

In 1865 my attention was again directed to the subject when considering Professor Holmes's magneto-electric light, and reference is made to this plan in Messrs. Stevenson's report of that date.[1] For such a purpose neither Holmes's nor Wilde's light could be employed, as they are produced by the rapid consumption of carbons, and require the employment of delicate lamp-machinery, which, though to a large extent automatic, involves the constant presence of a light-keeper in the lantern: so instead of these the electric spark without carbons was tried in the focus of optical apparatus. In order to increase the brightness of the flash, Rhumkorff's induction-coil was combined with a Leyden jar, on the suggestion of Professor Swan, who also advised the placing of the coil on the beacon, and the interrupter on the shore. A submarine cable was next procured by the Commissioners of Northern Lights. Mr. Hart designed an improved break, and Dr. Strethill Wright proposed the employment of two small coils instead of one large coil, in order to increase the volume of the flash, but the current failed to pass through the water as required.

In these circumstances, in Professor Swan's absence, Messrs. Stevenson applied to Dr. Siemens of London, the eminent electrician from whom the submarine cable was got, and he suggested another and very ingenious, though somewhat complicated, arrangement for producing the light, which he thus describes:—" The apparatus upon the beacon or buoy consists of a heavy electro-magnet, the coils of which are per-

[1] Report on the Application of the Magneto-Electric Light to Lighthouse Illumination. By D. and T. Stevenson. Edinb., November 1865.

manently connected with the conducting wire of the cable on the one hand, and with a contact-lever on the other hand, which contact-lever is actuated by the armature of the electro-magnet in the manner of a Neft's hammer. The circuit with the battery (consisting of from ten to twenty Bunsen's elements) on land is completed through the sea. When the current has had time to excite the electro-magnet sufficiently for it to attract its heavy armature, the motion of the latter breaks the circuit, which breakage is accompanied with a spark proportionate to the accumulated magnetism, and in some measure also to the capacity of the cable, which, in this apparatus, does not destroy but rather assists the effect. The luminous effect is increased by a slight combustion of mercury, which latter is continually renewed by a circulating pump worked by the armature, by which arrangement a good and permanent contact is insured."

This method was tried at Granton Harbour, and the current was passed easily through the wires, and the light produced by the deflagration of the mercury was very vivid. The result, however, was otherwise unsatisfactory, owing to the deposition of the products of combustion on the optical apparatus, and the existence of some mechanical difficulties which interfered with its continuous working. This instrument, therefore, which as respected the flash was very efficient, was still found inadequate for practical purposes.

Being thus again thrown back on the first plan of employing the induction-spark, the earth terminals were, on the advice of Professor P. G. Tait, increased in area, when the current was at once passed through the cable, and thus Professor Swan's original suggestion was successfully carried out. On this occasion the battery and break were placed on the eastern pier of Granton; and the submarine cable, which was upwards of half a mile long, extended to the Chain Pier at Trinity, on which were placed two induction-coils with condensers, and a holophotalised reflector to receive the spark. The battery consisted of sixteen Bunsen elements, and two additional cells were used for working the break. The coils contained each about four miles of wire. The spark produced was about quarter of an inch in length, bluish-white in colour, and very striking and characteristic in effect.

At the British Association meeting at Dundee in 1867, I stated that "the effect of the light might be also increased without using additional cells, if the same current could be again utilised so as to generate a second spark in the focus." This proposal may be carried out in one of two ways—*first*, by generating additional sparks in the same focus by using additional coils; or, *second*, by producing the separate sparks in the foci of separate reflectors contained in the lantern on the beacon, just as in some lighthouses several separate lamps are employed for increasing the effect. It may be useful to state that wires of different metals were tried for electrodes; and that, of all those tested, bismuth was found to give the brightest spark.

Trials were also made with the simple spark afforded by Wilde's electro-magnetic machine without carbons, and the result was highly satisfactory. The spark possessed much greater volume and power than with the Bunsen cells.

ILLUMINATION OF BEACONS AND BUOYS BY GAS.

6. As mentioned in the description of the Stornoway apparatus in the Transactions of the Royal Scottish Society of Arts, it occurred to me "that in some cases gas-pipes might be laid . . so as to illuminate a lantern placed on a beacon or buoy." At Stornoway, as already stated, all these plans were rejected in favour of the apparent light, from their liability to frequent extinction, which I regarded as an insurmountable objection, "at least in the present state of our knowledge"

The late Admiral Sheringham, in 1852, and therefore not very long after the erection of the Stornoway Beacon light, not only thought of applying gas to buoys, but made the experiment of illuminating one near Portsmouth, in 1853. Admiral Sheringham kindly sent me the following description of his interesting experiment:—

"My process was simply as follows:—From some convenient gas-works on shore I laid a series of ordinary iron pipes as far as possible, say to the low-water line. From thence I connected, by means of a union, a gutta percha tube; the one I used I think was about 2 inches in diameter, and perhaps about 50 or 60 fathoms long; this tubing was laid over the bank to the edge of the channel, where the buoy was placed for marking the limits of the fairway. The tube was carried up through the buoy, to about 3 or 4 feet above it, to the lanthorn provided with the necessary burners, etc.

"Copper wires were passed *through* the tube, *i.e.* inside of it, and carried over the gas-burner; here the wire was broken, and the two ends connected, at a sufficient distance apart, by means of platinum. The shore ends of the wire were in communication with a small galvanic battery, which, when in action, heated the platinum to a red heat. The gas was turned on by a tap on shore, and a most brilliant light instantly produced. It was quite marvellous to see the effect of its working; in one instant the light was put out and relighted.

"My experiments were made, I think, for three or four successive nights, at the entrance of Portsmouth Harbour, the gas being supplied from Mr. Hollingsworth's Assembly Rooms on the beach. The experiments were witnessed by many, and certainly were eminently successful, and were continued until a rascally barge parted from her anchors, drifted foul of it, and tore it all to pieces."

It is certainly to be regretted that these interesting experiments, in which it will be noticed Admiral Sheringham also used electricity, though not for the purpose of producing *light* but *heat*, were not prosecuted further; a fact which he informed me was not due to any doubt in the success of the system but owing to difficulties of a legal nature.

The first actual employment of gas for beacons was in 1861, on the Clyde, near Port-Glasgow, and the light has ever since been regularly maintained. It is situated about 300 feet from the shore, but instead of the method adopted by Admiral Sheringham for insuring the regular exhibition of the light, the supply of gas during the day is reduced, to a certain extent, by shutting a valve, and the full supply is let on at night. When the pressure of gas in the town of Port-Glasgow is reduced in the morning, the sinking of a float on the beacon shuts off the supply from the principal burner; while a small centre jet is kept lighted by means of a supply derived from a gasholder of the capacity of ten cubic feet, which is placed on the beacon, and is also connected with the main. From this source the small burner is lighted all day, and when the pressure is increased at night, the float which regulates the supply to the large burner rises, and thus opens the communication with the main. Owing to the fact of the river running close to the quay on the Port-Glasgow side, the fall on the main (which is of copper, and two inches in diameter) is landward from the beacon, so as to admit of a well for collecting the condensed water being placed at the shore end. The fact of this light having been now for nineteen years in existence goes far to prove that, in my first thoughts in 1851, the practical objections attending this method of illumination had been over-estimated for some localities.

The difficulties that are likely to attend the illumination by gas where buoys are used is the extinction of the light by the waves, and in all cases the accumulation of water in the main. If the extinction of the flame were noticed from the shore it would however be easy to re-light the gas by means of Admiral Sheringham's plan of heating a platinum wire.

There are many places, however, where these objections to the use of gas could be easily overcome, especially when the station is near the shore, as at Port-Glasgow, and when it is not exposed to a heavy sea.

PINTSCH'S GAS-ILLUMINATED BUOY.

This system of illuminating buoys promises to form a most important means of marking dangers by night, which have hitherto been altogether unprovided for. The gas-illuminated buoy is an ordinary one, of wrought-iron plates, about 60 feet capacity. It is gas and water tight, and capable of standing a pressure of 90 lbs. on the square inch. A pipe leads up from the top of the buoy to the lantern (Fig. 2, Plate XXIV.), which may be about 12 feet above the water. In the lantern, which is about 8 inches diameter, the gas-burner is placed, surrounded by a small fixed light apparatus. A Pintsch regulator, worked by a spring, is placed on the supply, so as to keep a constant pressure on the jet of $\frac{1}{2}$-inch of water. The buoy, which holds four months' supply, is charged by a boat coming alongside with a reservoir of the compressed oil-gas, which is made from shale. The cost of the light is stated to be from 3d. to 6d. per twenty-four hours.

Automatic Meter for producing Intermittent Lights by the flow of the Gas.—As regards the difficult problem of distinguishing one buoy from another, it is to be feared that it is not very practicable, and certainly far from desirable, to apply clockwork for producing rotation of the apparatus; and it occurred to me *to make the flow of the gas produce automatic intermittent action* by means of a dry meter. Such a form of meter must always pass a sufficient quantity of gas to secure the constant burning of a small jet, situate either immediately above or in the socket of a larger burner provided with a separate tube, for giving at regular intervals an increased supply which goes to the main burner, and is there ignited by the small jet. The full flame continues to burn until the action of the meter cuts off the larger supply and the small jet is again left burning alone. This process will of course go on continuously so long as the gas in the buoy is not exhausted.

I applied to Messrs. Milne, the well-known gas-engineers of Edinburgh, for their assistance in the matter; and they have, by altering the valves in a dry meter, made it to answer perfectly all the requirements. This mode of producing an intermittent effect was tested continuously for twenty-eight hours, which sufficiently proves its applicability to the production of intermittent light at ordinary lighthouses placed on the land, or at beacons, where a gasholder can be used. Actual experience is desirable in order to show whether it be equally applicable to floating lighted buoys, which are so much affected by the motion of the waves. If colour be adopted as an element of distinction, there may be obtained, by placing a lantern with coloured panes above another which is uncoloured, the following characteristics :—

Fixed Red and Green;
„ Red and White;
„ Green and White.

If, again, only one lantern be used, the following distinctions may be produced by the automatic meter :—

Revolving or Intermittent White;
„ „ Intermittent Red;
„ „ Intermittent Green.

Gymbals for preserving the Verticality of Buoy Light Apparatus.—Another difficulty which has been found to interfere with the proper exhibition of the light on floating buoys is the inclination of the buoy itself to the horizon, when acted on by strong winds and currents; and Mr. Cunningham, C.E., of Dundee, has designed a form of buoy which is expected to preserve its verticality in all states of the wind and sea. A much simpler mode of obviating the objection is to adopt gymbals, as employed in the Azimuthal Condensing Apparatus used in steamers' lights already described (Chap. III. **10**). The gas may be very simply conveyed to the burner by making the journals on which the lamp and apparatus swing hollow, so as to afford a passage for the gas.

Uniform System of Beacons and Buoys.

The earliest proposal to adopt a local buoyage system appears to have been that by Mr. R. Stevenson in his Report on the Improvement of the Forth Navigation in 1828, where black and red buoys and beacons were recommended for the different sides of the channel. This system is thus described in the report :—" For the use and guidance of river pilots, buoys and perches or beacons are likewise intended to be placed in the positions shown on the plan, those coloured red are to be taken on the starboard, and those coloured black on the larboard side in going up the river." This system has since been adopted by the General Lighthouse Boards in the United Kingdom.

In 1857 Admiral E. J. Bedford suggested a uniform system to be extended to all the coasts of the country, and this was introduced into Scotland by Mr. A. Cuningham, Secretary to the Lighthouse Board, and is still in use. The Trinity House also adopted a uniform but different system for the English coasts. In 1875 I suggested a simpler mode than any of these, but afterwards found it had been previously proposed (in 1859) by Mr. J. F. Campbell, "that a buoy shall indicate by its shape and colour the *compass direction* as well as the existence of a danger." The mode which I suggested for carrying out Mr. Campbell's system, which has since been adopted by the Norwegian Government, was the following:—

A Black Can Buoy will indicate dangers lying to the *westward* of the buoy.

A Red Nun Buoy will indicate dangers to the *eastward* of the buoy.

A Horizontally Striped Black and White Buoy, with a globe on its top, will indicate dangers to the *southward* of the buoy.

A Horizontally Striped Red and White Buoy, with a cross on the top, will indicate dangers to the *northward* of the buoy.

From these characteristics the following exceedingly simple sailing directions are obviously derived:—*Do not sail to the westward of a black can buoy; to the eastward of a red nun buoy; to the southward of a horizontally striped black and white buoy surmounted by a ball; or to the northward of a horizontally striped red and white buoy surmounted by a cross.*

Courtenay's Automatic Signal Buoy.

Plate XXIX. shows the Courtenay Buoy. A is a hollow cylinder of a length greater than the height of any wave that can occur in the locality where the buoy is to be moored; thus the surface level of the water within A remains stationary, while the buoy D with A attached rises and falls on passing waves. We have therefore the conditions of a fixed immovable column encompassed by a rising and falling envelope; in other words, we have a moving cylinder and a fixed piston, by which we can compress air by wave power.

The tube A extends to the top of the buoy, where a powerful whistle N is placed. E is a diaphragm in A, between which and the plate on the top of the buoy are two tubes G, open above, and having at their lower extremities ball valves F, as shown. A central open tube H leads from the diaphragm to the whistle. Suppose the apparatus to be carried from a position where the diaphragm E is just above the mean water level, that is the level of the column of water in A, to the summit of a wave, then the space between E and the top of the enclosed water column will have been greatly enlarged, and air must have been drawn in through the tubes G to fill it; now, as the instrument descends to the trough of the wave, the diaphragm must descend on the top of the water-piston, and the air compressed, being prevented by the ball valves F from escaping through the tubes G, is driven out through the central tube H, and so sounds the whistle N.

It is obvious that any disturbance of the surface of the water must produce this effect. Long low ground-swells must do it as well as short chopping waves; but of course the higher the wave the longer the duration of the sound. The buoy may be usefully employed wherever an undulation of 12 inches exists.

The buoy is fastened by a suitable anchor and chain attached to the mooring-shackle B, and is kept from any whirling motion by the rudder C. To guard against leakage, the tube I provides a passage from the space beneath the diaphragm E to that enclosed in the main shell. Compressed air is forced through this tube into the float, until the pressure in the latter equals the maximum pressure below the diaphragm. This tube is provided with a check-valve.

A Courtenay buoy, and also a common bell buoy (bell weighing 6 cwt.), were moored, in 1879, in 7½ fathoms of water, about 1⅓ mile off Inchkeith, in the Firth of Forth, in order that they might be tested as signals. The keeper at Inchkeith Lighthouse was directed to make observations on the action of both buoys, when it was found, after four months' trial, that the Courtenay buoy was heard on 108 days, while the bell buoy was only heard on 37 days. The keeper adds that the Courtenay buoy is "always the loudest and most continuous, attracting the attention of the listener much quicker than the bell buoy." It may also be remarked that the Courtenay buoy is sometimes heard at a distance of 5 miles.

General Remarks.—The experiments that have already been made on the application of electricity, and the facts which have been adduced regarding the present employment of gas, seem to show, so far as they go, the feasibility of the *submarine conduction* of other light-producing agency, and afford good hopes that by such means, and that of the Apparent Light, and Pintsch's gas buoy, an important system of coast-illumination may ultimately be introduced. The time is perhaps not far distant when the beacons and buoys in such a navigation as the entrance into Liverpool may be lit up by submarine conduction from a central station on either shore, while the whole management may be trusted to the charge of one or two light-keepers. The fact of a beacon having been illuminated for twenty-nine years on the Apparent Light principle at Stornoway, and another for nineteen years by gas conveyed under water at the Clyde, and the success which has attended the trials of Pintsch's illuminated buoys, seem sufficient to justify a farther and more general application of these principles to practice. The encouragements to proceed are many; and, to bring them more distinctly under view, I shall not only briefly recapitulate the various methods already described, and which afford very considerable scope for choice in applying them to the varying peculiarities of different localities, but will also adduce methods for the employment of sound, and other plans which have been elsewhere proposed.

I. *Different Sources of Illuminations for Beacons and Buoys.*

1*st*, The adoption of apparent lights.

2*d*, The use of dipping lights for indicating the position of shoals, so as to cover with the light the ground that is dangerous.

3*d*, The conduction either of voltaic, magnetic, or frictional electricity, or that produced by the efflux of steam, through wires, either submarine, or where practicable suspended in the air, so as to produce a spark either with or without vacuum tubes, or by means of an electro-magnet and the deflagration of mercury.

4*th*, The conduction of gas from the shore in submarine pipes.

5th, Self-acting electrical apparatus, produced by the action of sea-water or otherwise at the beacon itself, so as to require no connection with the shore. This method, which I suggested for beacons in the Society of Arts' Transactions for 1866, derives additional importance from the fact that the late Mr. A. Bain, the eminent electrician, informed me that in 1848 he produced a good light by using sea-water at Brighton.

6th, The illumination of buoys by compressed gas, invented and introduced by Mr. Pintsch.

II. *Different Applications of Sound for Warnings during Fogs.*

1st, The propagation of sound during fogs through pipes communicating with the shore,[1] or the origination of sound at the beacon or buoy itself, by compressing a column of air, or by acting on a column of water contained in the pipes.

2d, Bells rung by electricity. Mr. Wilde of Manchester, to whom I applied for his advice on this proposal, kindly informs me that, in his opinion, bells 12 or 18 inches diameter, placed on different beacons, and as far off as 10 miles from the shore, could be tolled a hundred times a minute by means of a 3½ or 4 inch electro-magnetic machine, worked by an engine of about two horse-power.

Mr. Mackintosh of Liverpool suggested, in 1860, that fog warnings might be made by a tide-mill, or a stream of land-water, for sending a current of air to work a pneumatic engine so as to ring a bell. Bells or whistles at a beacon may be also sounded, either by those means or by a hydraulic ram connected with a pipe, or by the rise and fall of the

[1] In one of the Paris water-pipes, 3120 feet long, M. Biot was able to keep up a conversation, in a very low tone, with a person at the other end.

tide lifting a weight on the shore, connected with a train of machinery on the beacon. The same moving power might also work an electro-magnetic, or other electric machine on the shore.

3*d,* The tolling of bells by the hydrostatic pressure of the tide, as proposed in 1810 by Mr. R. Stevenson for the Carr Rock. Professor Fleeming Jenkin has lately suggested to me the use of this method for working the contact-breaker for an electric current at the beacon, instead of on the shore, as in the experiments which have hitherto been tried.

Which of these many ways may be judged the best will depend on experience and the nature of each locality. The science of beacon illumination and identification is yet in its infancy, and we cannot limit its application.

In Plate XXVIII., Figs. 1, 2, 3, 4, and 5, show the buoys in general use on the coasts of Scotland; Fig. 6, a spar buoy used in the United States. Fig. 7 shows a proposal for reducing the force of the waves upon the buoy in very exposed situations, where one of the ordinary construction would not be likely to ride during storms. Plates XXX. and XXXI. show a malleable-iron and cast-iron beacon, originally used on the coast of Scotland.

CHAPTER VI.

ELECTRIC LIGHT.

THE history of the invention of the Electric Light and the modes of producing it are very fully given in Mr. Douglass's valuable paper in the Minutes of Proceedings of the Institution of Civil Engineers, vol. lvii. p. 77, from which are taken the following facts relating to the introduction of this most important luminary:—

The electric light, as employed in lighthouses, is primarily due to Faraday, who discovered that if a coil of wire were brought near to or drawn away from either of the poles of a permanent magnet, a current of electricity was induced. The first magneto-electric machine capable of generating a sufficient current for lighthouse purposes was made by Professor Holmes in 1853, and tried by the Trinity House at London in 1857; and in 1858 they first showed the electric light to the sailor. In 1847 Staite devised a lamp in which the upper carbon was made to approach the lower carbon by a current of electricity acting on an electro-magnet combined with clockwork; and Serrin appears to have produced the first complete automatic lamp for alternating currents. The Trinity House subsequently introduced the electric light at Dungeness in 1862, and at Souter in 1871.

The optical apparatus for these lights was designed by Mr. James T. Chance, and constructed by Messrs. Chance of Birmingham. Siemens and Wheatstone discovered in 1867 (but it is stated that Mr. Varley had made the same discovery in 1866), that powerful electric currents could be generated without a permanent magnet; and Holmes constructed, in 1869, for the Trinity House, a dynamo-electric machine on this new principle, which, according to photometric measurements by Dr. Tyndall and Mr. Douglass, produced a light of about 2800 candles.

The power of the apparatus at the high tower at the South Foreland is estimated at 152,000 candles, or twenty times that of the old dioptric oil light. Trials were made by Dr. Tyndall at the Lizard of various magneto-electric and dynamo-electric machines; and the results are given in the following Table:—

TABLE showing the COST, DIMENSIONS, WEIGHT, HORSE-POWER ABSORBED, and LIGHT PRODUCED, by the MAGNETO-ELECTRIC and DYNAMO-ELECTRIC MACHINES, tried at the SOUTH FORELAND, 1876-77.

NAME OF MACHINES.	Cost.	Dimensions.			Weights.	HP. absorbed.	Revolutions per Minute.	Light produced in Standard Candles.		Light produced per HP. in Standard Candles.		Size of Carbons.	Order of Merit.
		Length.	Breadth	Height.				Max.	Mean.	Max.	Mean.		
	£	Ft. in.	Ft. in.	Ft. in.	T. cwt. qr. lbs.							Inches.	
Holmes's Magneto-Electric	550	4 11	4 4	5 2	2 11 1 7	3·2	400	1,523	1,523	476	476	$\frac{3}{8}\times\frac{3}{8}$	6
Alliance do.	494	4 4	4 6	4 10	1 16 1 21	3·6	400	1,953	1,953	543	543	$\frac{3}{8}\times\frac{3}{8}$	5
Gramme Dynamo-Electric (No. 1) .	320	2 7	2 7	4 1	1 5 2 0	5·3	420	6,663	4,016	1,257	758	$\frac{1}{2}\times\frac{1}{2}$	4
Gramme do. (No. 2) .	320	2 7	2 7	4 1	1 5 2 0	5·74	420	6,663	4,016	1,257	758	$\frac{1}{2}\times\frac{1}{2}$	4
Siemens's do. (large) .	265	3 9	2 5	1 2	0 11 2 18	9·8	480	14,818	8,932	1,512	911	$\frac{11}{16}\times\frac{11}{16}$	3
Do. do. (small, No. 58 . . .	100	2 2	2 5	0 10	0 3 3 0	3·5	850	5,539	3,339	1.582	954	$\frac{1}{2}\times\frac{1}{4}$	2
Do. do. (small, No. 68 . . .	100	2 2	2 5	0 10	0 3 3 0	3·3	850	6,864	4,138	2,080	1,254	$\frac{1}{2}\times\frac{1}{2}$	1
Two Holmes's Magneto-Electric . .	1,100	9 10	4 4	5 2	5 2 2 14	6·5	400	2,811	2,811	432	432	$\frac{1}{2}\times\frac{1}{2}$	...
Two Gramme Dynamo .	640	5 2	2 7	4 1	2 11 0 0	10·5	420	11,396	6,869	1,085	654	$\frac{11}{16}\times\frac{11}{16}$	...
Two Siemens's do. (small, Nos. 58 and 68) .	200	4 4	2 5	0 10	0 7 2 0	6·6	850	14,134	8,520	2,141	1,291	$\frac{11}{16}\times\frac{11}{16}$	...

It is a fact of great significance that, owing to the progressive improvements that have taken place under the auspices of the Trinity House, the cost per unit of light produced has been reduced to 1-20th of what it was at Dungeness.

The French introduced the electric light at Cape La Hêve in 1863, and at Cape Grisnez in 1869. The optical apparatus for these lights was manufactured by Messrs. Sautter, Lemonnier, and Company of Paris. M. Sautter's results with the Gramme dynamo-electric machine, as reduced by Mr. Douglass to the usual standard of 9·5 candle power for a Carcel burner, show the intensity to be 2275 candles per horse-power for a 2½ horse-power machine, and 3221 candles per horse-power for a 13 horse-power machine. The effectiveness of this kind of apparatus is therefore greater than that of the Gramme and Siemens's machines at the South Foreland.

In 1866, Mr. Henry Wilde furnished for the Commissioners of Northern Lighthouses one of his electro-magnetic machines for producing the electric light, with which a series of experiments, about to be referred to, was made. Mr. Wilde kindly furnished me with a statement as to the origin of this invention. He says that "in 1864 I made the discovery that, when an electro-magnet was excited by the current from a magneto-electric machine, the magnetism of the electro-magnet was much greater than that of the entire series of steel magnets employed to generate the current. I also found that when an armature was rotated before the poles of the electro-magnet a current of much greater energy was produced from this armature than from that of the magneto-electric machine. Expressed generally, this discovery consists in developing from magnets and currents indefinitely weak, magnets and currents indefinitely strong.

"In the course of some experiments with a pair of electro-magnetic machines for producing electric light, I made the discovery that when two or more machines are driven by means of belts at about the same nominal speed, the combined currents so control the armatures as to cause them to rotate as synchronously as if they were mechanically geared together. As perfect synchronism of the rotations of the armatures is absolutely essential when the combined current of several machines is required, this electro-mechanical property of the current is an important one, as it allows of the machines being arranged without reference to their respective positions, or to their distances from each other."

The Merits of the Electric Light.—From the frequent occurrence of fogs on the coasts of this country, and the suddenness with which they come on, it is obviously of importance that all lights should be of as great penetrative power as possible. But there are certain headlands which, from their commanding geographical position, require more than others to be marked by lights of higher power. Although such sea-stations could, without doubt, be lighted up with oil so as to be made of great efficiency, yet if a better radiant than an oil flame can be got it ought to be applied at all those great salient points of the coast. Now the magneto-electric light, from its remarkable brilliance and splendour, presents the strongest claims. But the mere fact that a light has been discovered of far greater intensity than others, though important and encouraging, inasmuch as it justifies the most hopeful anticipations for the future, is not in itself sufficient to warrant its immediate general adoption. It is therefore essential that its application to lighthouse illumination should be fully tested. There are certain requirements for guiding the mariner, all of which must be fulfilled in the electric as in every other light.

1*st*, That it shall be constantly in sight during those periods of time at which it is advertised to the mariner as being visible.

2*d*, That it shall be seen in a thick and hazy atmosphere at the greatest possible distance.

3*d*, That it shall constantly maintain the distinctive character of the station where it is employed, so that it shall never be mistaken for any other light.

4*th*, That when the light is revolving its flashes shall remain long enough in view to enable the sailor to take the compass bearing of the lighthouse from the ship.

In considering how far these conditions are fulfilled by the electric light our attention must be directed to what at first sight appears to be its peculiar advantage, viz. its great power. To all near observers its brilliance and intensity are most striking and undeniable. But it has been and is still asserted, that at great distances the oil light maintains its power more fully than the electric. Such a phenomenon, certainly, seems to be very improbable; yet the subject in the present state of our knowledge is not one for dogmatic assertion, because it may be the case that the rays proceeding from the electric light suffer so much greater loss from absorption in passing through an obstructive medium than those from an oil flame, that the oil lights, if of equal initial power, may after all be the more powerful of the two at great distances. If this were really so, it would follow that the application of electricity to lighthouse illumination was based upon a fallacy. The mere glare or splendour of effect to a near observer, so far from being an advantage to the mariner, is a positive evil, and an evil which is submitted to only because it is believed to be a necessary accompaniment of the property of visibility at great distances.

All that the mariner can in any case require is *distinct visibility.* Anything short of this is useless; anything more than this is really mischievous, because by its lustre it tends to destroy his powers of perception of dangers in the water that are nearer his view, and which, therefore, from their proximity, threaten more immediate peril to the safety of his vessel. *The really useful power then, is that of penetration through an obstructing medium, and, therefore, the true measure of the usefulness of any light is the distance at which it remains distinctly visible, and preserves its characteristic appearance.*

The few observations made at Edinburgh seem rather to lead to the belief that the electric light suffers a greater amount of loss in penetrating the atmosphere than the oil light; but, as the original power is vastly greater, it will, nevertheless, be visible at much greater distances than an oil light.

The other qualities which have been specified are the uninterrupted exhibition of the light and the preservation of its distinctive character. There can be no doubt that the electric lights, both at Dungeness and Havre, especially when viewed at a distance, are liable to very great fluctuations in volume and intensity. Although such fluctuations are not likely, if of no greater duration than what was noticed, to lead to any doubt as to the identity of Dungeness or Cape la Hêve lighthouses, yet it would undoubtedly be of importance that the light should be rendered more certain in its exhibition, and less liable to variations in its power. It is well known that all lights are liable to twinkling in certain states of the atmosphere, but the cause which produces the flickering and momentary extinction of the electric light, though in many instances arising from imperfections in the quality of the carbons and from want of proper compensative action in the mechanism of the lamp, is also, undoubtedly, to a large extent due to the variation in the positions of the carbon points, and the consequent vertical variation of the plane of maximum intensity of the flame. This exfocal position of the flame must necessarily remove the beam of parallel rays from the horizon and throw it either too near the lighthouse or upwards to the sky.

While it is not to be disputed that the perfection of the electric lamp so as to ensure that the flame shall remain in a given constant position is the great problem which has to be solved, and to which Foucault, Duboscq, Holmes, and Serrin, have devoted much labour and ingenuity, there are, nevertheless, other means by which any irregularities in working the lamp may to a great extent be obviated. An important recommendation of the electric light, according to Professor Holmes, was that the apparatus may be held in a man's hat. But the electric light loses one of its greatest advantages by the employment of too small a size of optical apparatus, which, though less costly, is otherwise unsuitable, because every unit of surface in an apparatus of small size operates on a pencil of light of a greater angle of divergence than it would in larger apparatus. Hence, in the small apparatus, the grinding of the surfaces and fitting of the glass work become of the greatest importance, and must be executed in the most careful manner. Moreover, slight deviations from exact focusing will, with the small apparatus, produce greater deflections of the beam of parallel rays, and therefore those vertical displacements of the electric light which are constantly occurring, owing to the imperfection of the regulating machinery, will be especially injurious, because the angular displacement of the beam of emergent rays from the proper direction, for a given linear displacement of the carbon points, will *vary inversely in the simple ratio of the linear dimensions.* In addition to this, it may be remarked as a further disadvantage attending the use of small apparatus, that if a sufficient amount of divergence be obtained for the proper illumination of the ocean, this condition necessarily involves a corresponding divergence upwards upon the sky, so that a loss of light inevitably occurs. From what has been stated it will be seen that *the employment of small apparatus, such as was first adopted in England and France, for the purpose of producing vertical divergence upon the sea, is a retrograde movement.*

In order to take full advantage of the valuable properties of the electric spark, it is necessary that apparatus should be used of such a size as would, if made of the ordinary form, give practically speaking no divergence at all. For this purpose apparatus of the third order should, as first recommended by Messrs. Stevenson in their Report of 27th November 1865, to the Northern Lighthouses, be adopted in preference to the smaller orders. Mr. Chance states that he also suggested the same view to the Trinity House in 1862. For fixed lights the fourth order should for similar reasons be substituted for the sixth order. But the light proceeding from such apparatus, if made of the usual form, would of course be useless to the mariner, inasmuch as, in the case of a fixed light, no rays would be visible, excepting at or near the horizon; and so in the case of a revolving apparatus, in which there would be the further evil of the beam of light sweeping past the eye of the mariner so quickly as to prevent him from taking compass-bearings to the lighthouse. What is required then is to give the optical apparatus such a form as to produce the amount of *horizontal* divergence which is needed for taking compass-bearings, and also such an amount and direction of *vertical* divergence as will illuminate the sea from the horizon to the shore, and not waste any of the rays by illuminating the arc above the horizon. By such an arrangement alone will the mariner be enabled to reap the full benefit of the peculiar and valuable properties of this new radiant. In the recently erected lights of the Trinity House this has been most skilfully provided for by Mr. Chance (Plate XXXII. Fig. 4). The only cause for regret is that he should have adopted double agents from their being more easily constructed, but there is no one more able to overcome the difficulties of construction of single-acting differential apparatus than himself.

Experiments with the Electric Light.—Three different kinds of annular lens were tried at Edinburgh with Wilde's magneto-electric light in 1866. These instruments had each a focal distance of 500 millimetres, and were constructed of such forms as to produce different amounts of divergence. The experiments were thus made to include not only the trial of different forms of agent but also the determination of the amounts of horizontal and vertical divergence that are best suited for this radiant.

These instruments were, 1*st*, A plano-convex lens of the usual construction for revolving lights; 2*d*, A certain modification of the differential lens, the description of which was given in my replies to the scientific queries issued by the Royal Commission on Lighthouses in 1860.

Differential Lens.

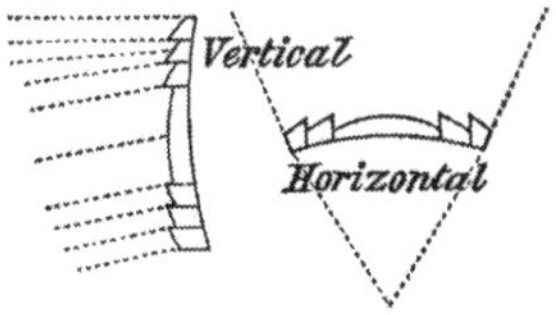

Vertical

Horizontal

Fig. 121.

The principle of this instrument is shown in Fig. 121, in horizontal and vertical section. Its outer face has the same profile as Fresnel's annular lens, but its inner face is ground so as to give different horizontal and vertical divergence. In the lens which was used in these experiments, however, it was judged better, in order to vary the results, to confine the curvature to the horizontal plane, so as to give a divergence in azimuth of 2½° (Plate XIX.); *3d*, The form of lens suggested by Mr. Brebner, C.E., which was divided into halves (Plate XIX.), and the rings of which by their construction and exfocal position, produced a divergence of 6° in azimuth, and 3° in altitude; or, in other words, it was calculated to give the *same horizontal and vertical divergence with the electric light as the common lens gives effectively with an oil light.*

Experiments were successively made with these three different instruments, with the following somewhat unlooked-for results:—When the frame revolved so as to give a flash every minute, the ordinary plano-convex lens was found, as had been expected, to be wholly unsuitable. The deficiency in horizontal divergence rendered it impossible to take a bearing to the light at the distance of 2½ miles, where the observers were placed; while the defect in vertical divergence was so great that the light was visible only for a very small distance above and below the focal plane. The horizontal divergence that had been given to the differential lens being somewhat less than half of that of a first-order revolving oil light, was found scarcely sufficient. The vertical divergence was of course insufficient as with the common lens. The double lens was found, as designed, to give in both planes about the same amount of effective divergence as a first-order oil light.

As regards power, the common plano-convex lens

proved of course the most powerful, the differential was somewhat less effective, but the double lens was not only very much inferior in power, but the distinguishing peculiarity of the electric flash was so far lost as to assimilate it in a great degree to an oil light. The penetrative powers of the instruments, as shown by the liquid ink-and-water photometer, were as under :—

Plano-convex lens, constructed to give parallel rays . . . . =1·00

Differential lens, constructed to give $2\frac{1}{2}°$ divergence in azimuth, and parallel rays in the vertical plane . . . . . . = ·90

Double lens, constructed to give 6° divergence in the horizontal, and 3° in the vertical plane = ·75

The results seem to show that the highly distinctive flash of the electric light, when acted on by optical apparatus, is not so much due to a greater amount of light as to the more complete parallelism of the rays arising from the smallness of the radiant. Apparatus of a small size, which necessarily produces a wasteful vertical divergence, should therefore be altogether discarded.

The general result of these investigations is, *First*, that the apparatus for revolving lights should be not less than the third order of 500$^{mm.}$ focal distance ; and, *Second*, that the annular lens should be either differential or of the double form. The differential may be either curved in both planes, or if only in the horizontal, then the carbons should be placed exfocally, as in the double lens.

There has lately been much discussion at the meetings of the Institution of Civil Engineers as to the relative penetrating powers of the electric and oil lights, and very different views have been expressed on the subject. But it appears to me that these discussions have been practically fruitless. The only mode of settling the question is to make observations at different distances on electric and other lights of *equal initial power*. The lights being equalised by observations near at hand, observations at greater distances would show whether they disappeared together at the same distance, or separately at different distances. Were this experiment tried, it would at once settle the vexed question of penetrability; but unless such experiments should prove that the electric light is very defective in penetrating power, which, from the practical success which has already attended its introduction seems very improbable, it must be regarded as immeasurably better fitted for lighthouse illumination than any other radiant, from its small divergence in optical apparatus. The actual practical results ascertained by M. Petit, chief of the Hydrographic service at Antwerp, on the South Foreland electric light, as compared with the oil lights of the North Foreland, Calais, Dunkerque, and Ostend, as stated by Dr. Tyndall, are, that "the first reaches its geographical range 75 times out of 100, while the second reaches its range only 29 times out of 100." But these facts, while proving the superior power of the South Foreland electric light, do not of course prove any superiority in specific penetrative power, as the amounts of light emitted from the different stations are not initially equal.

The following table of what he calls "focal compactness," has been computed by Mr. Douglass, and shows very clearly how much the electric surpasses all other illuminants in the ratio of intensity to area of flame :—

Comparative Focal Compactness of Lighthouse Luminaries.

Description of Luminary.	Diameter in the horizonal Focal Plane.	Area of Vertical Section.	Intensity. — Standard Candles.	Ratio of Vertical Area to Intensity. (Area Unity.)
	Inches.	Inches.		
OIL.				
Fresnel Lamps—				
1 wick	0·88	1·30	14	1 to 10·8
2 ,, ...	1·75	3·50	58	1 ,, 16·5
3 ,, ...	2·63	7·80	167	1 ,, 21·4
4 ,, ...	3·36	10·52	269	1 ,, 25·6
Improved Trinity House Lamps—				
1 wick	1·00	1·63	23	1 to 14·1
2 ,, ...	1·40	3·90	82	1 ,, 21·1
3 ,, ...	2·18	7·30	208	1 ,, 28·5
4 ,, ...	2·85	10·00	328	1 ,, 32·8
5 ,, ...	3·73	15·43	514	1 ,, 33·0
6 ,, ...	4·33	20·70	722	1 ,, 34·9
GAS.				
Wigham's System—				
28 jet burner	4·25	25·00	429	1 to 17·2
48 ,, ...	5·88	69·38	832	1 ,, 12·0
68 ,, ...	7·50	111·38	1,253	1 ,, 11·2
88 ,, ...	9·25	149·38	2,408	1 ,, 16·1
108 ,, ...	11·13	182·00	2,923	1 ,, 16·1
148 ,, ...	12·75	213·00	3,136	1 ,, 14·7
ELECTRICITY.				
Two Lizard Intensity—				
Minimum	0·72	0·41	8,250	1 to 20122·0
Maximum	0·96	0·72	16,500	1 ,, 22916·6
Mean	0·84	0·57	12,375	1 ,, 21519·0

CHAPTER VII.

STATISTICS OF LIGHTHOUSE APPARATUS.

(Extracted from M. Allard's *Mémoire sur l'Intensité et la Portée des Phares:* Paris, 1876.)

THE following Chapter contains the principal results of the very valuable researches of M. E. Allard, by whose kind permission it has been compiled.

Dimensions of Wicks and Burners, etc.

Order of Apparatus.	Diameter of Apparatus.	Number of concentric Wicks.	Outer Diameter of Wicks.	Space between Wicks.	Mean Diameter of Wicks.	Height of Button above Burner.	Sum of lengths of unfolded Wicks in each Burner.
1st Order	1·84 m.	5	11 cm.	5 mm.	10·5 cm.	21 mm.	1021 mm.
2d „	1·40 „	4	9 „	„	8·5 „	19 „	691 „
3d „	1·00 „	3	7 „	„	6·5 „	17 „	424 „
4th „	0·50 „	2	5 „	„	4·5 „	15 „	220 „
5th „	less than 0·50 m.	1	3 „	„	2·5 „	13 „	78 „

Consumption of Oil in relation to Diameter of Burner.

If c denote the consumption of mineral oil in grammes per hour, d the diameter of the burner in centimètres, then

$$c = 4{\cdot}9 d^{2{\cdot}22}.$$

Height and Volume of Flame.

If h and d denote the height and diameter in centimètres, s the apparent surface, and v the volume of a flame, then $h = 2{\cdot}73\sqrt{d}$; $s = \frac{1}{4}\pi hd = 2{\cdot}144 d^{\frac{3}{2}}$; $v = \frac{1}{6}\pi d^2 h = 1{\cdot}4294 d^{\frac{5}{2}}$.

Volume of Flame in relation to Consumption of Oil for the different Burners.

If u denote the volume of flame in cubic centimètres per gramme of oil consumed in an hour, d the diameter of the burner in centimètres, then

$$u = 0 \cdot 2917 d^{0 \cdot 28}.$$

Luminous Intensities.

A carcel burner consuming 40 grammes of colza oil per hour being taken as unity, if I denote the intensity for mineral oil in a burner of diameter d, then

$$\mathrm{I} = 0 \cdot 22 d^{2 \cdot 1}.$$

The Transparency of Flames.

Considering a horizontal cylinder of flame of length l, if the flame were perfectly transparent the intensity of light would be directly proportional to l, but in reality for any value a, of the coefficient of transparency there is a limiting intensity which may be called the intensity for $l = \infty$. Now for $a = \cdot 7$ the value of l to give an intensity only $\frac{1}{10}$ less than the limiting intensity is 2·58 inches; and for $a = \cdot 8$ the corresponding value of l is 4·13 inches: thus it appears that an intensity differing little from the limiting intensity can be obtained from a flame of no great thickness.

Coefficient of Transparency.

The fraction of the whole incident light which passes through a sheet of flame one centimètre (about $\frac{2}{5}$ of an inch) thick $= \frac{4}{5}$.

Consumption of Oil in relation to Intensity of Flame.

Number of wicks to the burner	1	2	3	4	5	6
Consumption of oil per unit of intensity if flames were perfectly transparent (in grammes per hour) . .	19·75	17·27	15·50	14·18	13·07	12·24
Do. per unit of intensity actually obtained . .	24·93	25·20	25·87	26·81	27·75	28·96

Luminous Intensities of Apparatus; Loss due to Reflection, Absorption, and Framing of Apparatus.

The loss due to surface *reflection* on entering and leaving the glass may be valued at ·050, ·052, ·058, ·075, ·120, ·230, for angles of incidence respectively = 0°, 15°, 30°, 45°, 60°, 75°.

In totally reflecting prisms the luminous ray suffers three deviations instead of two, therefore the above values should be multiplied by $\frac{3}{2}$.[1]

The loss by *absorption* in the glass, although properly given by an exponential formula, may, with sufficient accuracy, be taken as ·03 per centimètre of glass traversed.

The loss due to the horizontal *joints* of the lenses, and to the *intervals between the reflecting rings*, varies from ·02 to ·03, or from ·01 to ·04, in passing from the 1st to the 5th Order.

[1] This assumes the loss at the reflecting side to be the same as that at the refracting sides, which does not quadrate with Professor Potter's observations given in Chapter II.

Total loss due to these three causes for fixed lights :—

		Angle occupied by each part.	Loss.
1st Order	Upper rings . . .	From 29°·2 to 76° . . .	·30
	Drum	2 × 28°	·13
	Lower rings . . .	From 30°·2 to 48°·5 . . .	·30
2d Order	Upper rings . . .	From 31°·6 to 79°·6 . . .	·30
	Drum	2 × 31°·3	·13
	Lower rings . . .	From 32°·7 to 51°·6 . . .	·30
3d Order	Upper rings . . .	From 35°·1 to 82°·1 . . .	·29
	Drum	2 × 33°·3	·13
	Lower rings . . .	From 36° to 54°·2 . . .	·29
4th Order	Upper rings . . .	From 34°·1 to 80°·1 . . .	·29
	Drum	2 × 30°·8	·13
	Lower rings . . .	From 34°·8 to 63°·4 . . .	·29
5th Order	Upper rings . . .	From 34°·1 to 80°·6 . . .	·27
	Drum	2 × 30°·8	·13
	Lower rings . . .	From 34°·8 to 62°·8 . . .	·27

Theoretical quantities of light emitted in an angle of 1° contained between two vertical planes, when losses are calculated as above :—

1st Order.				2d Order.	3d Order.	4th Order.	5th Order.
Upper Rings.	Drum.	Lower Rings.	Total.	Total.	Total.	Total.	Total.
573	1623	181	2377	1653	981	464	147
Corresponding quantities photometrically observed to be emitted.							
572·5	1620	182·5	2375	1650	977·5	464	146·5
Observed axial intensities.							
225b	760b	105b	1090b	600b	280b	74b	17·5b

b represents the intensity of a Carcel burner consuming 40 grammes of colza oil per hour.

For *electric lights* the axial intensities lie between 16,200 and 5600. The upper, middle, and lower parts of a fixed light apparatus produce intensities of light which are in the ratios 2, 7, and 1, where 10 is that of the whole apparatus.

Coefficients of the Different Fixed Light Apparatus.

These are the ratios in which the intensity of the lamp is increased by the apparatus. Where m is the coefficient, f the focal distance, d the diameter, and h the height of flame, they can be calculated from the formula—

$$m = \tfrac{2}{3}\left(\frac{f}{\sqrt{d}}\right)^{1\cdot15}, \text{ or } m = 2\cdot12\left(\frac{f}{h}\right)^{1\cdot15}.$$

Order of apparatus	1st	2d	3d	4th	5th
Intensity of lamp	36^b	24^b	$14{\cdot}3^b$	$6{\cdot}9^b$	$2{\cdot}2^b$
Coefficient m obtained from photometric observations .	30·28	25·00	19·58	10·72	7·96
Coefficient m obtained from the formula	30·47	24·95	19.58	10·71	7·98

Intensities of Annular Lenses.

Suppose the light from a 1st order lens subtending an angle of 43°·3 in the horizontal plane to be received on a screen at a suitable distance, an inverted image of the flame will be obtained on the screen. The following Table shows the variation of intensity in this image on moving from the axis 1° at a time in the vertical and horizontal planes :—

Degrees.	3°	2°	1°	0°	1°	2°	3°	Sum.	
								Total.	For 1°.
2°	...	70^b	230^b	280^b	230^b	70^b	...	880^b	20^b
1°	370^b	1550	2220	2460	2220	1550	370^b	10740	248
0°	1630	4700	6550	7150	6550	4700	1630	32910	760
1°	750	2280	3200	3530	3200	2280	750	15990	369
2°	...	950	1560	1770	1560	950	...	6790	157
3°	...	250	810	910	810	250	...	3030	70
						Totals	. .	70340	1624

The accuracy of the observations in the horizontal focal plane may be tested by adding the intensities obtained from degree to degree and dividing their sum by 43°·3 the angle of the lens, when the quotient ought to be the same as the axial intensity found for the drum of the corresponding 1st order fixed light. The sum is 32910, and this divided by 43·3 gives 760 (see observed axial intensities, p. 195). Hence a relation can be established between the intensity of a fixed light and that of the corresponding flashing light. For if a straight line through the focus be taken as axis of abscissæ, and from it ordinates be drawn representing the intensities in the focal plane, the curve obtained by joining the extremities of these ordinates approximates to a parabola. Hence if A denote the axial intensity of the flash, or the distance from focus to apex, y the intensity at another point in the focal plane situated at x degrees from the axis, α the horizontal semi-divergence, we have the relation—

$$y = \mathrm{A}\left(1 - \frac{x^2}{\alpha^2}\right)$$

The sum of the intensities or the quantity of light corresponding will be represented by the surface of this parabola, which $= \frac{4}{3}\,\mathrm{A}\alpha$. On the other hand, if a be the intensity of the fixed light, and ϕ the angle of the lens, the quantity of light in this angle will be ϕa, and it must equal that which the lens concentrates into the angle of horizontal divergence; hence $\frac{4}{3}\,\mathrm{A}\alpha = \phi a$, and $\mathrm{A} = a\,\frac{3\phi}{4\alpha}$. Thus the intensity of an annular lens is obtained by multiplying that of the corresponding fixed light by $\frac{3}{4}\,\frac{\phi}{\alpha}$, where ϕ is the angle subtended by the annular lens, and α the horizontal semi-divergence.

Axial Intensities of the different Lenticular Apparatus.

Order of Apparatus.		Angle subtended by the Lens.	Mean Horizontal Divergence 2α.			Coefficient $\frac{3\phi}{4\alpha}$			Intensity of the Fixed Light.				Intensity of the Flash.				*Ratio* of Intensity of Flash to that of Fixed Light.
			Upper Prisms.	Drum.	Lower Prisms.	Upper Prisms.	Drum.	Lower Prisms.	Upper Prisms.	Drum.	Lower Prisms.	Total.*	Upper Prisms.	Drum.	Lower Prisms.	Total.*	
1st Order 6 Wicks	$\frac{1}{8}$	43°·3	6°·4	7°·7	6°·7	10·15	8·43	9·69	285[b]	963[b]	134[b]	1382[b]	2893[b]	8118[b]	1298[b]	12,309[b]	8·9
	$\frac{1}{12}$	28·2				6·61	5·49	6·31					1884	5287	846	8,017	5·8
	$\frac{1}{16}$	20·9				4·90	4·07	4·68					1394	3919	627	5,940	4·3
	$\frac{1}{24}$	13·5				3·17	2·63	3·02					903	2533	405	3,841	2·8
1st Order 5 Wicks	$\frac{1}{8}$	43·3	5·4	6·5	5·7	12·03	9·99	11·40	226	765	106	1097	2719	7642	1208	11,569	10·5
	$\frac{1}{12}$	28·2				7·83	6·51	7·42					1770	4980	787	7,537	6·87
	$\frac{1}{16}$	20·9				5·81	4·82	5·50					1313	3687	583	5,583	5·09
	$\frac{1}{24}$	13·5				3·75	3·12	3·55					848	2387	376	3,611	3·29
1st Order 4 Wicks	$\frac{1}{8}$	43·3	4·4	5·3	4·7	14·76	12·25	13·82	169	571	80	820	2494	6995	1106	10,595	12·92
	$\frac{1}{24}$	13·5				4·60	3·82	4·31					777	2181	345	3,303	4·03
2d Order 5 Wicks	$\frac{1}{8}$	42·8	7·0	8·4	7·3	9·17	7·64	8·79	160	574	66	800	1467	4385	580	6,432	8·04
	$\frac{1}{12}$	27·6				5·91	4·93	5·67					946	2830	374	4,150	5·2
	$\frac{1}{16}$	20·6				4·41	3·68	4·23					706	2112	279	3,097	3·87
	$\frac{1}{20}$	15·9				3·41	2·84	3·27					546	1630	216	2,392	2·99
2d Order 4 Wicks	$\frac{1}{8}$	42·8	5·7	6·9	6·0	11·26	9·30	10·70	120	430	49	599	1351	4000	524	5,875	9·8
	$\frac{1}{12}$	27·6				7·26	6·00	6·90					871	2580	338	3,789	6·3
	$\frac{1}{16}$	20·6				5·42	4·48	5·15					650	1926	252	2,828	4·7
	$\frac{1}{20}$	15·9				4·18	3·46	3·98					502	1488	195	2,185	3·65
2d Order 3 Wicks	$\frac{1}{8}$	42·8	4·3	5·3	4·6	14·93	12·11	13·96	83	295	34	412	1239	3572	475	5,286	12·8
	$\frac{1}{20}$	15·9				5·55	4·50	5·18					461	1328	176	1,965	4·77

3d Order 4 Wicks	$\frac{1}{6}$	57·5	7·7	9·6	8·4	11·20	8·98	10·27	84	291	32	407	941	2613	329	3,883	9·5
	$\frac{1}{8}$	42·3				8·24	6·61	7·55					692	1924	242	2,858	7·02
	$\frac{1}{10}$	33·2				6·47	5·19	5·93					543	1510	190	2,243	5·5
	$\frac{1}{12}$	27·2				5·30	4·25	4·86					445	1237	156	1,838	4·5
	$\frac{1}{16}$	19·7				3·84	3·08	3·52					323	896	113	1,332	3·27
3d Order 3 Wicks	$\frac{1}{6}$	57·5	6·0	7·5	6·5	14·38	11·50	13·27	58	200	22	280	834	2300	292	3,426	12·2
	$\frac{1}{8}$	42·3				10·58	8·46	9·76					614	1692	215	2,521	9·0
	$\frac{1}{10}$	33·2				8·30	6·64	7·66					481	1328	169	1,978	7·07
	$\frac{1}{12}$	27·2				6·80	5·44	6·28					394	1088	138	1,620	5·78
	$\frac{1}{16}$	19·7				4·93	3·94	4·55					286	788	100	1,174	4·19
3d Order 2[illegible] Wicks	$\frac{1}{6}$	57·5	4·3	5·4	4·6	20·05	15·97	18·75	34	117	13	164	682	1868	244	2,794	17·0
	$\frac{1}{16}$	19·7				6·87	5·47	6·42					234	640	83	957	5·8
4th Order 3 Wicks	$\frac{1}{6}$	56·0	13·2	15·1	14·1	6·36	5·56	5·96	26	90	10	126	165	500	60	725	5·75
	$\frac{1}{10}$	31·7				3·60	3·15	3·37					94	284	34	412	3·27
4th Order 2 Wicks	$\frac{1}{6}$	56·0	9·4	10·8	10·1	8·94	7·78	8·32	15·3	52·8	5·9	74	137	411	49	597	8·07
	$\frac{1}{10}$	31·7				5·06	4·40	4·71					77	232	28	337	4·55
	$\frac{1}{12}$	25·7				4·10	3·57	3·81					63	188	22	273	3·7
5th Order 1 Wick	$\frac{1}{6}$	56·0	5·6	6·5	6·1	15·00	12·92	13·84	6·6	22·9	2·5	32	99	296	35	430	13·4
	$\frac{1}{10}$	31·7				8·49	7·32	7·80					56	162	19	237	7·4
5th Order ·375 m. 2 Wicks	$\frac{1}{2}$	165·0	12·5	14·4	13·4	19·80	17·19	18·47	10·6	38·2	4·2	53	210	657	78	945	17·8
	$\frac{1}{6}$	55·0				6·60	5·73	6·16					70	219	26	315	5·94
	$\frac{1}{10}$	31·0				3·72	3·23	3·47					39	123	15	177	3·34
5th Order ·375 m. 1 Wick	$\frac{1}{2}$	165·0	7·5	8·6	8·1	33·00	28·78	30·56	4·6	16·6	1·8	23	152	478	55	685	29·8
	$\frac{1}{6}$	55·0				11·00	9·59	10·18					51	159	18	228	9·9
	$\frac{1}{10}$	31·0				6·20	5·41	5·74					29	90	10	129	5·6
5th Order ·3 m. 1 Wick	$\frac{1}{6}$	55·0	9·4	10·8	10·1	8·33	7·64	8·17	3·5	12·6	1·4	17·5	29	96	11	136	7·7

* For loss in passing through panes of lantern and other causes, $\frac{1}{5}$ may be deducted from the above total intensities.

Dioptric Spherical Mirror.

The intensity of light from an ordinary fixed light apparatus is increased 38 per cent in the angle corresponding to the mirror.

Increase of the Height of the Drum.

An angle of 38° instead of 30° above the axis should be subtended by the dioptric lens. In one respect it would be advantageous to increase the angle farther; for the absorption of light due to thickness of glass is greater in the rings than in the lenses; but, on the other hand, it is necessary to take into account the effects due to coloured dispersion. In the elements of the lens, the dispersions which are produced at the entrance and exit of the luminous ray are added because they are in the same direction, and their sum increases with the distance of the horizontal axis; the last elements of these lenses give very marked colorations, which cause a small loss of luminous intensity. In the catadioptric rings, on the other hand, the internal reflection causes no dispersion; the refractions at entrance and exit produce two dispersions nearly equal, and in opposite directions, so that emerging rays have no sensible colour. Hence in this respect it would be disadvantageous to increase much the vertical amplitude of the dioptric lens. Taking into account all the aspects of the question, 38°, or at most 40°, may be taken as a reasonable limit.

Obstruction of Light by the Burner.

In order to diminish the obstruction of luminous rays going in the direction of the lower parts of the apparatus, each wick of the burner has been made to stand above that which comes next it in passing from the centre outwards.

Absorption of Light by the Atmosphere.

If a denote the coefficient of transparency, that is the proportion of light which unit length of the medium allows to pass, L the intensity which the light from the source would have at unit distance in a vacuum, and y the intensity of this light at distance x in the absorbing medium, then the law of intensity in terms of the distance will be expressed by

$$y = \mathrm{L}\frac{a^x}{x^2}$$

Taking the kilomètre as unit of distance, the value of a obtained from Bouguer's experiments in clear air is $\cdot 973$; while in a fog at Paris in 1861 the value of a was reduced to $(\cdot 62)^{1000}$, or $\cdot 62$ was the fraction of the whole light allowed to pass by a layer of air 1 mètre thick.

Equation of Luminous Ranges.

For every observer there is a limit of luminous intensity λ, below which the eye is no longer affected. Putting

$$\mathrm{L}\frac{a^x}{x^2} = \lambda$$

the value of x obtained shows the range of the light L. For a person with average eyesight, λ may be taken $= \cdot 01$ of a Carcel burner.

State of the Atmosphere defined by the range of the Unit of Light.

If p be the range in mètres of the unit of light, its value may be found from the equation $a^p = p^2\lambda$;

and for $a = \cdot 973$, $p = 8860$ mètres.

„ „ $a = (\cdot 62)^{1000}$, $p = 25$ „

Here it must be noted that p has different values for different eyesights.

In the Table (Plate XXXIII.), the left hand column contains values of luminous intensity in Carcel burners; the lower column contains different values of a, the coefficient of transparency of the atmosphere. To each of the coefficients a there corresponds a value of the range of the unit intensity, and these ranges are inscribed along the upper column of the table, while the oblique lines represent luminous ranges. The use of the table thus composed will be easily understood, thus: if one desires to know the range of a light for a given value of a, he looks in the column to the left for the value of its intensity, 100 for example; at the same time he looks in the lower column for the value of a, ·871 for example, and, on ascending vertically to the horizontal line through 100, he arrives at a point of intersection through which passes the oblique line marked 22; the range sought for is therefore 22 kilomètres. These three numbers, 100, ·871, and 22, are so connected that any one of them may be found by means of the other two. When the point of intersection of the horizontal and vertical lines is not traversed by an oblique line, the aid of the eye must be called in to determine the range required. In the table given in M. Allard's *Mémoire*, each side of the small squares here given is subdivided into five parts, so that closer values of the intensity and transparency are given; the table being large enough to deal with intensities varying from ·32 to 10,000,000 burners, while he shows more extended values of a.

Another useful table given in M. Allard's *Mémoire* may be mentioned. It is divided into three parts. The first gives the luminous ranges corresponding to different intensities of light in times of dense fog. It shows that when the unit Carcel burner can penetrate only 27¼ yards, it would require an intensity 45 times greater than that of the strongest electric light used in France (Gris-Nez) to penetrate twice that distance, or 54½ yards. Again, when the unit intensity can penetrate 3270 yards, it would require an intensity 1½ time that of the Gris-Nez light to penetrate four times that distance.

The second part gives intensities required to carry from ·27 to 162 nautic miles in states of the atmosphere called foggy, mean, and clear, for which the unit intensity can penetrate respectively from 5341 to 7085 yards, from 7085 to 7630 yards, and 9347 yards.

The third shows ranges, corresponding to different intensities, from ·1 to 10,000,000 Carcel burners in the same three states of the atmosphere.

CHAPTER VIII.

SOURCES OF ILLUMINATION.

THE application of the electric spark to lighthouse illumination having been dealt with in Chapter VI., the present will be confined to the consideration of the other sources of light employed for that purpose.

Fig. 122 is a section of the Argand burner, in which *t* is the cylindrical space through the centre of the wick for the passage of the second air current, and *c* the glass chimney. It is an inch in diameter, and the central air space is $\frac{3}{4}$ inch diameter. Fig. 25 (page 60) shows the arrangement of the Argand lamp used in the Northern Lighthouse Service.

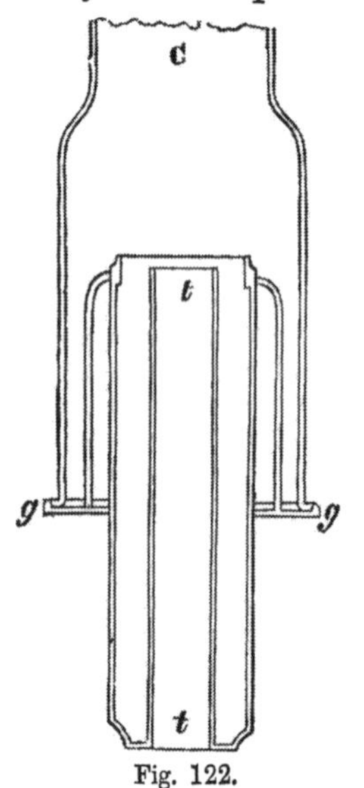

Fig. 122.

Fresnel Burner.—When Fresnel designed his dioptric apparatus he at once saw the necessity for a greater intensity of flame than could be got from the Argand, and, with the assistance of Arago and Mathieu, he devised multiple wick burners, previously suggested by Rumford; Figs. 123 and 124 are plan and section of one of his burners, with four concentric wicks, the spaces between them being passages which allow air to pass up to the inner wicks; *C*, *C′*, *C″*, *C‴*, are the rack handles for raising and lowering the wicks; *A B* is one of the ducts

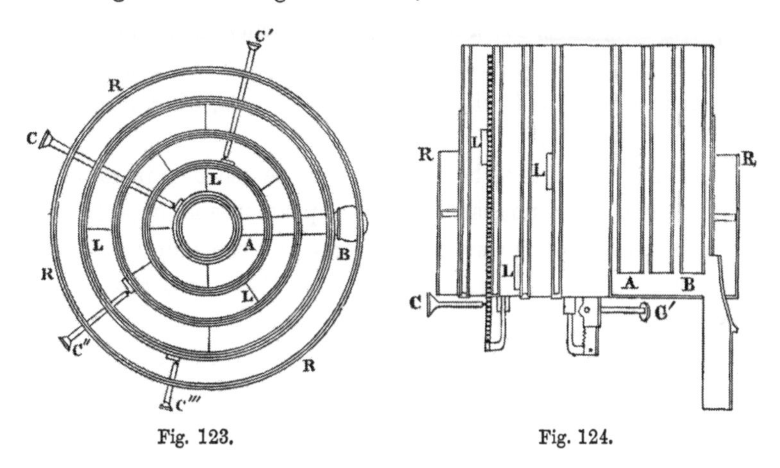

Fig. 123. Fig. 124.

which lead oil to the four wicks; *L*, *L*, *L*, are small plates of tin by which the burners are soldered together, and placed so as not to hinder the free passage of air to the flame. The air spaces are about ¼ inch wide. The chimney is carried by the gallery *R R*, and is surmounted by a sheet-iron cylinder, which serves to give it a greater length, and has a small damper fixed in it, capable of being turned by a handle for regulating the current of air. In order to keep the burner cool and protect it from the excessive heat produced, the oil is supplied in superabundant quantity, so as to overflow the wicks. Fresnel used two, three, and four concentric wicks; Mr. Alan Stevenson, in 1843, introduced a fifth; and Mr. Douglass has more recently adopted a sixth.

It will be observed that the leading principle of this, as of all other lighthouse burners introduced since Argand's time, whether for burning vegetable, animal, or mineral oils, is his double current.

The Argand and Fresnel burners were constructed for burning animal and vegetable oils, those most employed being sperm, lard, olive, cocoa-nut, and colza, which last was used by the majority of lighthouse services since the middle of the present century till within the last few years, when mineral oil took its place.

The advantages to be derived from the use of mineral oil as a lighthouse luminant, were so apparent and important, viz. increased luminous intensity for equal consumption, and a cost per gallon one-half that of colza, that many attempts were made to devise a burner for its consumption.

This, however, was found no easy matter, as the proportion of carbon to hydrogen in mineral oils is very large, as compared with that in animal or vegetable oils, and consequently the mineral oils require much greater quantities of air for their combustion.

Maris Burner.—In 1856 M. Maris devised a single-wick burner, with a disc or deflector at the top of the central passage, which had the effect of throwing the central current of air more into the flame. This burner gave good results with mineral oils, and was employed by the French Lighthouse authorities in several of their harbour lights.

Doty Burner.—All attempts however, to burn mineral oils, either in the Fresnel or specially constructed multiple wick burners, failed till so recently as 1868, when Captain Doty solved the problem and brought out his mineral oil burner. Fig. 125 is a section and plan of a four-wick Doty burner. By a happy choice of proportions in the various parts of his burner, and by the addition of an exterior cylinder surrounding the outer wick, and a central disc, both placed in such a manner as to throw a current of air into the flame at the right place, Doty succeeded in producing single and multiple wick hydro-carbon burners, which carry a flame of great luminous intensity and regularity. The overflow of oil required to keep the colza burners cool is unnecessary in these, and therefore the oil is maintained at constant level, this being effected in the ordinary mechanical and moderator lamps by an ingenious stand pipe, devised by Doty, and represented in Fig. 125. The following table gives the details of these burners, and also the candle power and consumption, as determined by Dr. Stevenson Macadam—

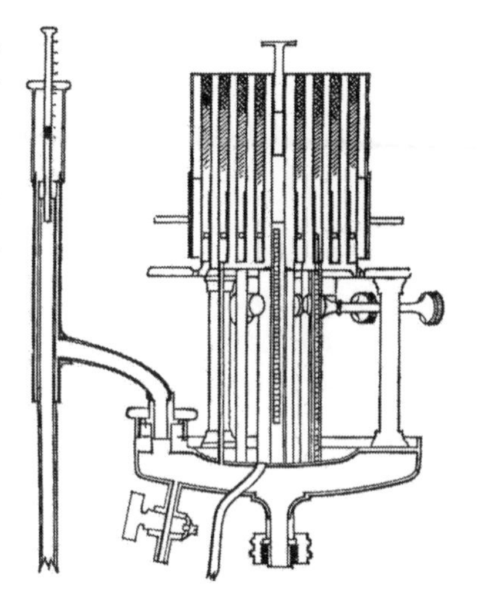

Fig. 125.

Nos. of wicks.	Mean diameter of outer wick in inches.	Height of intense part of flame excluding tails, in inches.	Candle powers in sperm candles, each consuming 120 grs. sperm per hour.	Value of light from consumption of 1 gallon in lbs. sperm.	Consumption per hour in gallons.
1	·82	1	23·65	27·39	·0148
2	1·75	1½	80·13	27·04	·0508
3	2·5	1¾	200·75	27·2	·1262
4	3·2	2	287·62	27·3	·1801

Captain Doty first of all brought his burners under the

notice of the French Lighthouse authorities, and they tried them in a first-order light in December 1868; in 1870 they converted all their third and fourth orders to paraffin; and in 1873 they resolved to adopt his burners in all classes of lights.

In September 1870 the Commissioners of Northern Lights tried Doty's four-wick burner at Girdleness, in Scotland, one of their first-order lights; and in March 1871 they approved of a report by Messrs. Stevenson, their engineers, recommending their general adoption in all classes of lights in the service, and this recommendation has since been carried out.

These burners have also been introduced, under the direction of Messrs. Stevenson, into the lighthouse services of China, Japan, New Zealand, and Newfoundland, and experience has fully justified their claim to efficiency, durability, and simplicity; as a proof of the real importance of the change from colza to mineral oils from an economical point of view, it may be observed that, in the case of the Scotch lighthouses, an annual saving of between £4000 and £5000 has been effected. It may be stated that the Doty burners are equally suitable for the consumption of colza oil.

The Trinity House did not adopt the Doty burner; but their engineer, Mr. J. N. Douglass, introduced the one shown in Fig. 126, and its action will be readily understood from what has been said in connection with other burners. In it Mr. Douglass employs an outside jacket, a cone-shaped perforated disc or button, and also tips for the tops of the wick cases, which can readily be taken off and replaced when burned out; an eventuality, however, which does not occur in the Doty burners. By means also of an interior deflector, which can be put on when required, and by igniting only half the number of wicks (the outer three in a six-wick for instance), the power of the burner can be diminished by one-half. This is Mr. Douglass's lamp of single and double power, for use in clear and foggy weathers respectively, and by it the "flexibility" claimed for gas is to some extent introduced into oil lamps. Mr. Douglass claims to have obtained even better results with this burner than those of Doty stated above. It is right to state that Capt. Doty claims the Trinity burner as being the same as his own.

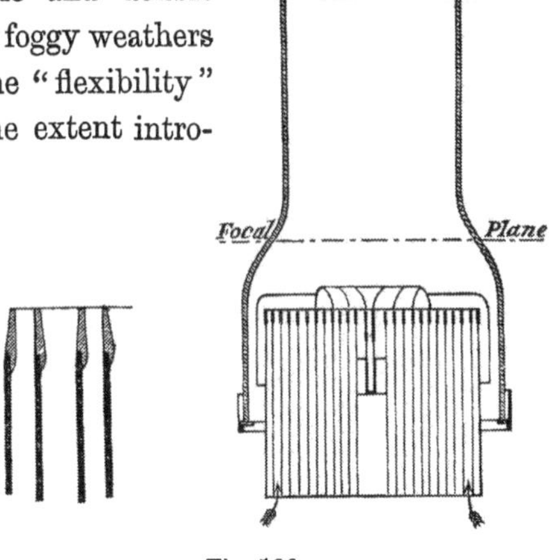

Fig. 126.

Messrs. Farquhar, Silber, Funk, and others, have also produced burners, but their use is very limited so far as lighthouse illumination is concerned.

Paraffin Oil.—Mineral oil is found in a natural state in wells, principally in America, and is also distilled from bituminous schist or shale on a large scale in France, Germany, and Great Britain. That employed most extensively in lighthouse illumination is Scotch paraffin, distilled from shale found in the county of Linlithgow, and is generally considered the best for the purpose. It gives a flame of great luminous intensity; is quite safe, and low in price. The specific gravity, which is a test of the relative richness of the oil, should be from 0·8 to 0·82 (water = 1) at 60° Fahr., and the flashing point or temperature at which it begins to evolve inflammable vapour should not be lower than 125° Fahr., nor higher than 135° Fahr.

Fountain Lamp.—Single-wick burners are supplied with oil from the cistern or fountain (which may be placed at one side, or immediately below them) by the capillary action of the wick alone. But in the case of multiple wick burners other methods must be employed to secure a sufficient supply.

Mechanical and Moderator Lamps.—If the cistern be placed above the level of the top of the burner the flow of the oil to the wick cases is effected by the direct action of gravity, regulated by a contrivance which maintains a constant head. If, however, the cistern be placed below this level, either a *mechanical lamp* is employed, in which the oil is forced into the burner by pumps worked by clockwork, or a *moderator* lamp, in which this is effected by the pressure exerted by a weighted piston descending in a cylinder forming the cistern.

Coal Gas.—Coal gas was first used as a lighthouse illuminant at Salvore, near Trieste, in 1817. For many years it has been used in the harbour lights of Great Britain when in the neighbourhood of gasworks, and the burners employed were either those giving an ordinary flat flame, or, in some cases, a circular flame on Argand's principle. Recently Mr. J. R. Wigham, of Dublin, has designed a new form of gas burner, which the Commissioners of Irish Lights have adopted in several of their lighthouses.

Mr. Wigham's compound or crocus burner consists of a group of 28 vertical tubes, each carrying an ordinary double fish-tail burner. When the gas issuing from all these jets is ignited, the incandescent gases unite into one large flame, and by suspending, immediately above it, a chimney of talc, which not only induces a powerful current of air but also acts as an oxidiser, the excess of carbon at the top of the flame is rendered incandescent (adding to the intensity of the flame), and is ultimately consumed. By means of a mercurial valve, which acts both as a connection and regulating gas cock, additional groups, of 20 jets each, can readily be

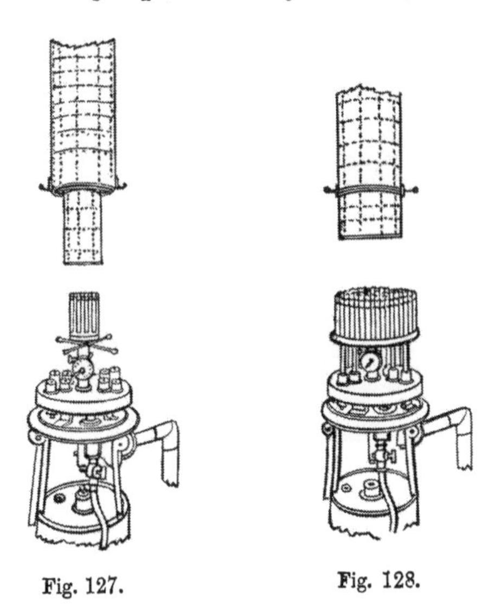

Fig. 127. Fig. 128.

arranged round the first, which forms a central nucleus; and in this way, depending on the state of the atmosphere, the power of the burner can be made at will 28, 48, 68, 88, or 108 jets. Fig. 127 is an elevation of the burner, as arranged for 28 jets, with its oxidiser above. Fig. 128 is the same for 108 jets, and Fig. 129 is an enlarged section of the 108-jet burner. In what he calls his biform, triform, or quadriform systems, referred to in Chap. X., Mr. Wigham places two, three, or four of the burners already described vertically one above the other, and by means of an iron casing or flue the products of the combustion of the lower burners are intercepted and thrown outwards, so as not to interfere with the upper burners, while pure air is supplied to each of them by cylindrical openings brought through the flues.

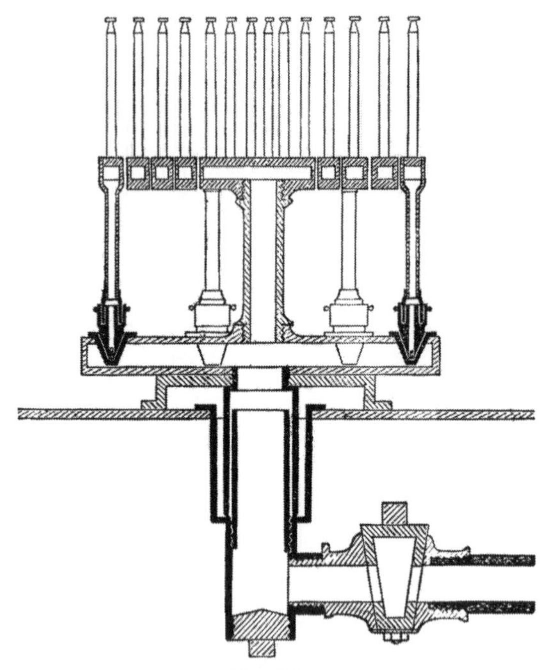

Fig. 129.

The following table gives the candle powers, etc.—

Nos. of Jets.	Consumption of Cannel Gas per hour.	Candle power in Sperm Candles, consuming 120 grs. per hour.
28	51·4	429·6
48	93·2	832·0
68	146·3	1250·18
88	244·0	2408·0
108	308·0	2923·0

The diameter of the 108-jet burner is 12 inches.

Among the many other sources of light which have been proposed for lighthouse illumination may be mentioned Drummond's oxyhydrogen light; Gurney's proposal to supply pure oxygen to an oil flame; and Mr. Wigham's method of introducing the electric spark, or oxyhydrogen light, into the centre of his gas flame.

CHAPTER IX.

MISCELLANEOUS SUBJECTS CONNECTED WITH LIGHTHOUSES.

MR. ALAN STEVENSON ON THE DISTRIBUTION OF LIGHTS ON A COAST.

"THE considerations," says Mr. Stevenson, in 1848, "which enter into the choice of the position and character of the lights on a line of coast, are either, on the one hand, so simple and self-evident, as scarcely to admit of being stated in a general form, without becoming mere truisms; or are, on the other hand, so very numerous, and often so complicated, as scarcely to be susceptible of compression into any general laws. I shall not, therefore, do more than very briefly allude to a few of the chief considerations which should guide us in the selection of the sites and characteristic appearance of the lighthouses to be placed on a line of coast. Perhaps these views may be most conveniently stated in the form of distinct propositions:—

"1. The most prominent points of a line of coast, or those first made on *over-sea* voyages, should be first lighted; and the most powerful lights should be adapted to them, so that they may be discovered by the mariner as long as possible before his reaching land.

"2. So far as is consistent with a due attention to distinction, revolving lights of some description, which are necessarily more powerful than fixed lights, should be employed at the outposts on a line of coast.

"3. Lights of precisely identical character and appearance should not, if possible, occur within a less distance than 100 miles of each other on the same line of coast which is made by over-sea vessels.

"4. In all cases the distinction of colour should never be adopted except from absolute necessity.

"5. Fixed lights and others of less power may be more readily adopted in narrow seas, because the *range* of the lights in such situations is generally less than that of open sea-lights.

"6. In narrow seas also the distance between lights of the same appearance may often be safely reduced within much lower limits than is desirable for the greater sea-lights. Thus there are many instances in which the distance separating lights of the same character need not exceed 50 miles; and peculiar cases occur in which even a much less separation between similar lights may be sufficient.

"7. Lights intended to guard vessels from reefs, shoals, or other dangers, should in every case where it is practicable be placed *seaward* of the danger itself, as it is desirable that seamen be enabled to *make* the lights with confidence.

"8. Views of economy in the first cost of a lighthouse should never be permitted to interfere with placing it in the best possible position; and when funds are deficient, it will generally be found that the wise course is to delay the work until a sum shall have been obtained sufficient for the erection of the lighthouse on the best site.

"9. The elevation of the lantern above the sea should not, if possible, for sea-lights, exceed 200 feet; and about 150 feet is sufficient, under almost any circumstances, to give the range which is required. Lights placed on high headlands are subject to be frequently wrapped in fog, and are often thereby rendered useless, at times when lights on a lower level might be perfectly efficient. But this rule must not, and indeed cannot be, strictly followed, especially on the British coast, where there are so many projecting cliffs, which, while they subject the lights placed on them to occasional obscuration by fog, would also entirely and permanently hide from view lights placed on the lower land adjoining them. In such cases, all that can be done is carefully to weigh all the circumstances of the locality, and choose that site for the lighthouse which seems to afford the greatest balance of advantage to navigation. As might be expected in questions of this kind, the opinions of the most experienced persons are often very conflicting, according to the value which is set on the various elements which enter into the inquiry.

"10. The best position for a sea-light ought rarely to be neglected for the sake of the more immediate benefit of some neighbouring port, however important or influential; and the interests of navigation, as well as the true welfare of the port itself, will generally be much better served by placing the sea-light *where it ought to be*, and adding, on a smaller scale, such subsidiary lights as the channel leading to the entrance of the port may require.

"11. It may be held as a general maxim that the fewer lights that can be employed in the illumination of a coast the better, not only on the score of economy but also of real efficiency. Every light needlessly erected may, in certain circumstances, become a source of confusion to the mariner; and, in the event of another light being required in the neighbourhood, it becomes a *deduction* from the means of distinguishing it from the lights which existed previous to its establishment. By the needless erection of a new lighthouse, therefore, we not only expend public treasure but waste the means of distinction among the neighbouring lights.

"12. Distinctions of lights, founded upon the minute estimation of intervals of time between flashes, and especially on the measurement of the duration of light and dark periods, are less satisfactory to the great majority of coasting seamen, and are more liable to derangement by atmospheric changes, than those distinctions which are founded on what may more properly be called the *characteristic appearance* of the lights, in which the times for the recurrence of certain appearances differ so widely from each other as not to require for their detection any very minute observation in a stormy night. Thus, for example, flashing lights of five seconds' interval, and revolving lights of half-a-minute, one minute, and two minutes, are much more characteristic than those which are distinguished from each other by intervals varying according to a slower series of 5″, 10″, 20″, 40″, etc.

"13. Harbour and local lights, which have a circumscribed range, should generally be fixed instead of revolving; and may often, for the same reason, be safely distinguished by coloured media. In many cases also, where they are to serve as guides into a narrow channel, the leading lights which are used should, at the same time, be so arranged as to serve for a distinction from any neighbouring lights.

"14. Floating lights, which are very expensive, and more or less uncertain from their liability to drift from their moorings, as well as defective in power, should never be employed to indicate a turning-point in a navigation in any situation where the conjunction of lights on the shore can be applied at any reasonable expense."

MR. ALAN STEVENSON'S TABLE OF RANGES OF LIGHTS.

TABLE of DISTANCES at which OBJECTS can be SEEN AT SEA, according to their respective elevations, and the elevation of the eye of the Observer.

Heights in Feet.	Distances in Geographical or Nautical Miles.	Heights in Feet.	Distances in Geographical or Nautical Miles.	Heights in Feet.	Distances in Geographical or Nautical Miles.
5	2·565	70	9·598	250	18·14
10	3·628	75	9·935	300	19·87
15	4·443	80	10·26	350	21·46
20	5·130	85	10·57	400	22·9
25	5·736	90	10·88	450	24·33
30	6·283	95	11·18	500	25·65
35	6·787	100	11·47	550	26·90
40	7·255	110	12·03	600	28·10
45	7·696	120	12·56	650	29·25
50	8·112	130	13·08	700	30·28
55	8·509	140	13·57	800	32·45
60	8·886	150	14·22	900	34·54
65	9·249	200	16·22	1000	36·28

Example.—A tower 200 feet high will be visible to an observer whose eye is elevated 15 feet above the water, 20·66 nautical miles; thus, from the table :—

15 feet elevation, distance visible 4·44 nautical miles.
200 „ „ „ 16·22 „ „

20·66

FLOATING LIGHTS.

Prior to the year 1807 the only kind of floating light was a ship with small lanterns suspended from the yardarms or frames. The late Mr. Robert Stevenson realised the inutility of such a mode of exhibition, and conceived the idea of forming a lantern to surround the mast of the vessel, and to be capable of being lowered down to the deck to be trimmed, and raised when required to be exhibited. His plan had the advantage of giving a lantern of much greater size, because it encased the mast of the ship, and with this increase of size it enabled larger and more perfect apparatus to be introduced, as well as gearing for working a revolving light. Fig. 130 shows this lantern, and the following is his description of it:—

"The lanterns were so constructed as to clasp round the masts and traverse upon them. This was effected by constructing them with a tube of copper in the centre, capable of receiving the mast, and through which it passed. The lanterns were first completely formed, and fitted with brass flanges; they were then cut longitudinally asunder, which conveniently admitted of their being screwed together on the masts after the vessel was fully equipped and moored at her station. Letters *a a* show part of one of the masts, *b* one of the tackle-hooks for raising and lowering the lanterns, *c c* the brass flanges with

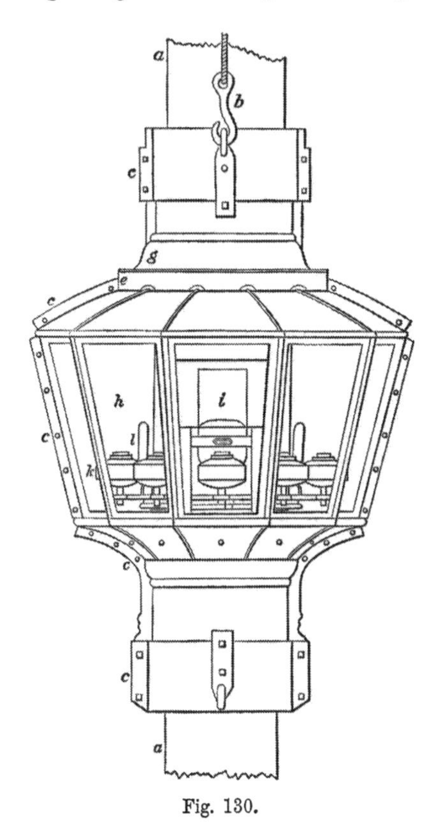

Fig. 130.

their screw-bolts, by which the body or case of the lantern was ultimately put together. There were holes in the bottom and also at the top connected with the ventilation; the collar-pieces *e* and *g* form guards against the effects of the weather. The letter *h* shows the front of the lantern, which was glazed with plate-glass; *i* is one of the glass shutters by which the lamps were trimmed—the lower half being raised, slides into a groove made for its reception; *k* shows the range of ten agitable burners or lamps out of which the oil cannot be spilt by the rolling motion of the ship. Each lamp had a silvered copper reflector *l*, placed behind the flame."[1]

After that important alteration, *revolving* catoptric apparatus was applied to floating lights in England; and M. Letourneau, in 1851, proposed to employ different fixed *dioptric* apparatus in one lantern for floating lights.

Plate XXVI. shows elevation and horizontal section of one of the Hooghly Floating Lights on the dioptric principle, designed for the Indian Government by Messrs. Stevenson. It will be observed that four of the separate lights are always visible in every azimuth. Similar floating lights have since been constructed for the Government of Japan. The Hooghly lights were seen for a distance of 19 miles, and the first of these was reported by the officiating attendant master to the secretary of the Government of Bengal "to be a most efficient one, and the best and most brilliant that has ever been exhibited from a light-vessel in these seas."

Plate XXVII. shows a section of one of the latest light-ships of the Trinity House, which are on the *catoptric* principle, kindly communicated by Mr. J. N. Douglass. A steam siren fog-signal apparatus and windlass are shown on the draw-

[1] Account of Bell Rock Lighthouse. Edinburgh, 1824.

ing. Mr. Douglass informs me that during the last five years the Trinity House have been replacing the 6-feet octagonal lanterns with cylindrical ones 8 feet diameter; and 21-inch diameter reflectors have also been substituted for 12-inch. The "Seven Stones" lightship, which has all the recent Trinity House improvements, is moored off the Land's End in 42 fathoms. Her moorings consist of a 40-cwt. mushroom anchor, and 315 fathoms of 1½-inch unstudded chain cable. The vessel, which was specially designed by Mr. Bernard Waymouth, secretary to Lloyd's, is timber-built, copper-fastened throughout, sheathed with Muntz metal, and of the following dimensions, viz.—length, 103 feet between perpendiculars; extreme breadth, 21 feet 3 inches; depth of hold from the strake next the timbers to the upper side of the upper deck beams, 10 feet 3 inches. In the event of the vessel breaking adrift (a most improbable occurrence), she is provided with foresail, lug mainsail, and mizzen, the latter being frequently used with advantage for steadying the vessel at her moorings. The cost of the vessel, fully equipped for sea, with illuminating and fog signal apparatus complete, was about £9500.[1]

LANDMARKS.

Where there is a sunk rock which is so low in the water as to preclude the erection of a beacon, towers are sometimes erected on the land for guiding vessels clear of the danger, but these guides are limited to the azimuthal plane. It is worthy of consideration whether such indications of the place of danger might not be extended to the *vertical* plane, at places where the land is suitable and the rock is not very far from the shore.

[1] Min. of Proc. of Inst. of Civil Engineers, vol. lxii.

Fig. 131 represents in plan and section a headland, with rocks lying both near the shore and farther out in the sea.

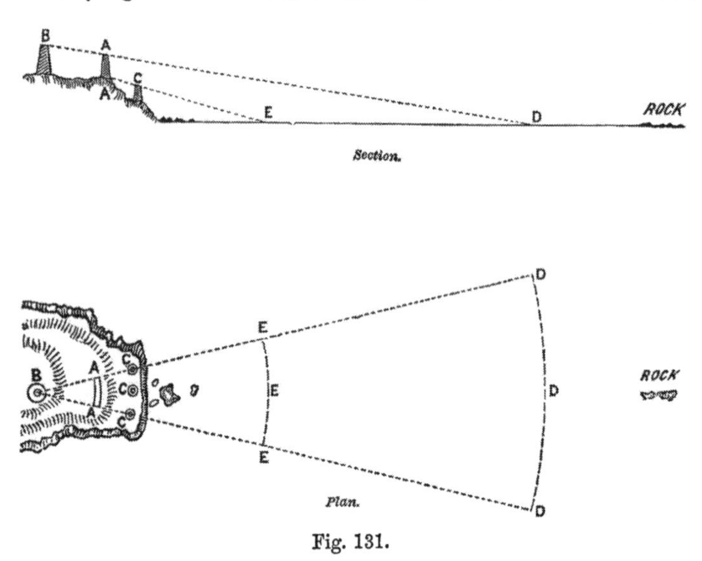

Fig. 131.

In order to mark off the safe limits of the channel, a wall A, circular in the horizontal plane, might be built on the headland, and at its centre of curvature a tower B, of greater height than the wall, and also three low towers, C C C, between the wall and the sea. An imaginary line joining the tower with the *top* of the wall would mark off the seaward limit of safety, that would keep vessels sufficiently clear of the outer rock. Another imaginary line, A C E, joining the *bottom* of the wall with the tops of the low towers would in like manner point to the verge of safety on the side next the land. The sailor would therefore never steer so far from the land as to see the top of the high tower projecting above the top of the wall, nor so near the shore as to see the top of any of the low towers projecting above the *bottom* of the wall. Thus, by keeping the high tower always *out of view*, and the bottom of the wall always *in view* (clear of the low towers), he would never be in danger. For greater distinction when snow is on the ground, the upper part of the high tower should be painted red, the wall black with white stripes, and the low towers white.

LANTERNS.

The lantern of a lighthouse consists of a glazed framing supporting a dome and ventilator. Its object is to protect the light and apparatus from wind and rain, and yet give free ventilation inside, not only to cause the lamp to burn well, but also to prevent the condensation of moisture on the inside, and consequent clouding of the panes of glass.

The early form of lantern had vertical astragals, which necessarily reduced the amount of light in the particular arcs which they subtended.

Mr. Robert Stevenson improved the dome by introducing the present inner dome, the air space between the two acting as a bad conductor of heat, and preventing the rapid condensation of water that would otherwise take place with a single dome. Mr. Alan Stevenson, as already mentioned in Chapter II., introduced inclined or diagonal framing for the lantern, as shown on Plate XI., thus not only making the lantern much stronger to resist any rocking motion, but also causing the loss of light by the astragals to be spread nearly uniformly over the whole lighted arc. The astragals in the Northern Lighthouse service are made of gun metal (having a tensile strength of 33,000 lbs. to the square inch), glazed with flat plate-glass held in its place by means of gun-metal strips screwed into the astragals. The dome is made of copper plates riveted together, and the following is the specification of the glass of a first-order lantern—"The panes are to be of the best quality of polished mirror plate-glass, not less than $\frac{1}{4}$ inch in thickness, nor more than $\frac{5}{16}$ inch. The glass requires to be perfectly polished, to have truly parallel faces, to be free from air bubbles, colour, and striæ, or other blemishes, and to be carefully rounded on the edges by grinding, after being cut to the proper size."

Mr. Douglass's lantern, Plate XII., already referred to, also shuts out only a small amount of light, and spreads the loss equally. This lantern is cylindrical, the diagonals are made helical, and of steel, and the glass is curved so that the light may fall on it radially. The panes are $\frac{5}{8}$ inch thick.

In France, and also in America, the lanterns are still made with vertical astragals, and are glazed with glass $\frac{5}{16}$ inch thick, protected, in some situations, from breakage by birds being driven against them, by means of brass wire gauze, the meshes being about $\frac{1}{4}$ inch square.

The lanterns of *apparent lights* should be well ventilated, so as to prevent a haze from forming on the inside of the lantern glass.

Glazing of Lanterns.—Great care is bestowed on the glazing of the lantern, in order that it may be quite impervious to water during the heaviest gales. There is a certain amount of risk of the glass plates being broken by the shaking of the lantern during high winds; and to prevent this as much as possible various precautions are adopted. The arris of each plate is always carefully rounded by grinding; and grooves, about $\frac{1}{2}$ inch wide, capable of holding a good thickness of putty, are provided in the astragals for receiving the glass, which is $\frac{1}{4}$ inch thick. Small pieces of lead or cork are inserted between the frames and the plates of glass, against which they may press, and by which they are completely separated from the more unyielding material of which the lantern frames are composed.

Storm Panes.—In the event of a breakage occurring to any of the panes in the lantern, which is not unfrequently the case from birds or small stones being driven against it during

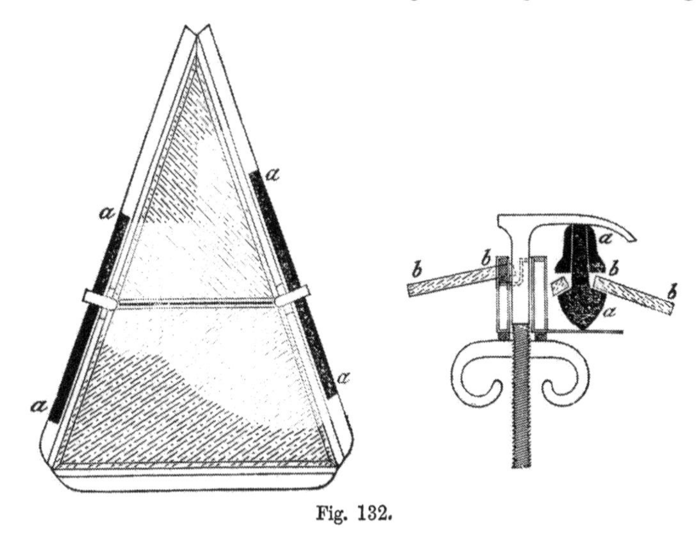

Fig. 132.

gales, glazed copper frames, as shown on Fig. 132, are used in the Northern Lighthouse service. They are fixed to the astragals from the inside by means of the screws shown in the woodcut. *a a* are the astragals, *b b* are the glass panes.

Lightning-Conductors.

As lighthouse towers generally occupy exposed situations on high headlands or isolated rocks, they are peculiarly liable to be struck by lightning, and the question of their protection demands the special attention of the engineer, for an efficient lightning-conductor may be said to be an essential part of every properly designed lighthouse establishment.

Franklin, about the year 1749, conceived the idea that electricity, which is produced in the laboratory, and lightning were identical; and founding on this, and his previous discovery of the action of metallic points on electricity, he published a pamphlet in 1750, in which he states that all damage done to buildings by electricity descending from the clouds, could be prevented, by fixing on their summits iron rods with sharp points, the lower ends of the rods being fixed in moist ground. In 1752 he made his celebrated experiment with a kite, demonstrating the truth of his former theories, and the same year he erected a lightning-conductor for the protection of his own house.

At the present time the question of protection from lightning rests nearly in the same state as that in which Franklin left it; the chief improvement in construction being the substitution of copper instead of iron as the material for the conductor, and although the effects of lightning are well known to all, and much has been written on the subject, science has still, as in Franklin's day, to confess its inability to advance any perfectly satisfactory theory either of the genesis or subsequent development of thunderstorms.

Although the practical requirements of a good conductor were discovered and fully enunciated by Franklin, those who came after him fell, in many cases, into serious errors, and more harm than good resulted from the use of their conductors. It may be said that till comparatively recently, the original requirements of Franklin were in practice generally neglected. These requirements are:—

1*st*, A terminal point.

2*d*, A good earth connection; and

3*d*, A continuous metallic connection between them.

The action of this arrangement is supposed to be as

follows:—A thunder-cloud, inducing an equal and opposite charge on the earth beneath it, approaches a building so protected, and while still at too great a distance for a disruptive charge to take place between the cloud and the building, or "to strike it," as it is called, the pointed rod, in virtue of the well-known action of points, begins to act by silently and continuously draining off from the earth a sufficient amount of electricity to render the cloud perfectly innocuous.

Copper being, of all metals, the best conductor of electricity, is therefore the most suitable material for a lightning rod, and may be used for this purpose either in the form of round rods, flat bars, or wire rope, care being taken that the metal is tolerably pure, and that there is metallic contact at the joints, which in the case of rods is effected by screwed coupling boxes; and, though less satisfactorily, with a half check and screws in the case of bars. The use of wire rope secures more perfect continuity, and being very pliable it has the advantage of being more easily trained and secured to the building.

The rods should be carried about 18 inches above the highest part of the lantern, and there terminated in two or three branches, each being finished with a sharp but not slender point, and fitted with a platinum cap. The object of having more than one point is to reduce the risk of the metal of a single point being fused. The use of platinum in forming the fine point is chiefly to preserve it from oxidation. The conductor is carried down the outside of the tower, from the top to the bottom, as a continuous rod, and carefully put in metallic connection with all large masses of metal inside or outside the building, such as machinery, bells, stoves, pipes, tanks, ladders, etc. This precaution was adopted by Robert Stevenson in his design for the Bell Rock Lighthouse, after consultation with Playfair, Leslie, and Brewster. Professor Tait was also recently consulted by the Commissioners of Northern Lighthouses. But besides being continuous, the conductor must be of sufficient section to carry any charge to which it may be subjected without fusing, and in most situations a $\frac{3}{4}$ inch rod or wire rope, as recommended by Faraday, will be found quite sufficient. The rod should be attached to the masonry by copper bats, at every 10 feet apart; and in towers exposed to the stroke of the sea the bar should be sunk in a raglet in the tower, to prevent its being disturbed by the waves.

On reaching the ground it should be carried in a trench to a spot 20 or 30 feet from the foundation of the tower, and there divided into two or three branches, to each of which a copper plate should be soldered, and these in turn buried in the earth or sunk in water. The size of these copper plates depends on the nature of the soil in which they are placed, but $20'' \times 12'' \times \frac{1}{8}''$, is the size generally adopted in the Northern Lighthouse service. A "good earth terminal" is of the very greatest importance, moist earth or large bodies of water being the most favourable, and it is at this point that the lighthouse engineer, especially at exposed rock stations, meets with the greatest practical difficulty. For although nothing can be better than the sea as the earth connection, yet on a rocky face, exposed to the rise and fall of the tide and the full action of the waves, it must be admitted to be in certain situations hardly practicable to provide that even for a single winter's storms the termination of the conductor, however well secured, shall remain so undisturbed as to insure its constant immersion in the water during all states of the sea and tide.

A tower, furnished with a lightning rod as described above,

constructed with ordinary care and occasionally inspected ocularly, will be quite safe from demolition and even serious damage. But to prevent the possibility of even slight damage it would be necessary periodically to test electrically the whole rod from the upper point to the earth. This can be effected by employing an apparatus, several varieties of which have recently been brought out, the principle involved in all of them being to measure the resistance of the conductor to the passage of a current of electricity generated in a small portable battery, by means of a series of resistance coils and a galvanometer. Testing, however, as described above, has not as yet come into general use in this country, although employed to some extent in Germany.

RADIAL MASKING SCREENS.

Figs. 133 and 134 show in elevation and plan radial masking screens, designed by the late Mr. J. M. Balfour, for

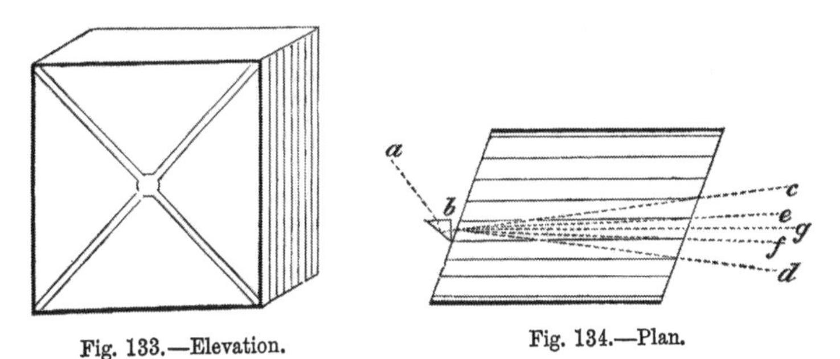

Fig. 133.—Elevation. Fig. 134.—Plan.

cutting off sharply, in a given direction, the light proceeding from any apparatus. These screens, which are in use at several of the Northern Lighthouses where the fairway for ships is narrow, intercept all rays of light which have too great divergence, and prevent them from passing beyond the line in which the light has to be cut off, while they permit the parallel rays to pass through, as shown in Fig. 134, where *b c* and *b d* are the rays which are intercepted, *a* is the focal point, *b* a reflecting prism, while the rays *b e* and *b f* pass freely through.

MACHINERY FOR REVOLVING LIGHTS.

The machines employed for causing the frame to revolve were originally of comparatively light clockwork. When the dioptric system was introduced a somewhat heavier construction of machinery was adopted; and when the holophotal arrangement was introduced, the much greater mass of glass, with its heavier armature, required a still more substantial and powerful class of wheels and pinions. The steel friction rollers required for dioptric lights have also, in like manner, been increased in number from 12 in the original apparatus to 16, which is the number adopted for the Mull of Galloway Condensing Intermittent light. Mr. David Stevenson, in 1849, enlarged the driving-wheel at North Ronaldshay holophotal apparatus, which at first had an irregular staggering motion, to 4 feet 8¾ inches diameter, which has been found to produce a smooth and steady revolution. Instead of an outside driving-wheel an internal wheel was adopted, which might be regarded as a circular rack, 5 feet in diameter, worked by an internal pinion; so that, as the teeth worked internally, nothing could accidentally be brought in contact with the gearing, the outer periphery of the driving-wheel having a smooth surface, and the pinion working inside instead of outside. For small class apparatus for Japan, he had also covered the peripheries of the wheels with bands of indiarubber, which, simply by their friction, secured a much smoother and steadier motion than that produced by toothed gearing. Mr. Douglass, in 1871, enlarged the diameter of the driving wheel to that of the revolving frame.

ASEISMATIC ARRANGEMENT.

A less common source of danger than any of those already dealt with, to which a lighthouse may be exposed, is that of earthquakes, which are believed to be the result of a sudden blow delivered to the crust of the earth by the collapse of internal cavities, the generation of steam, or the sudden fracture of large masses of rock previously strained. From the locus of this blow a series of concentric waves of vibration is propagated through the crust of the earth to the buildings situated on the surface, and the direction of the motion there produced will of course depend on the direction in which the wave is travelling when it arrives there.

The motion will therefore be truly vertical at a place on the earth's surface immediately above the seat of the disturbance, but will acquire a greater and greater horizontal component, as the place is farther and farther removed from the points on the earth's surface where the motion is vertical.

The action to which buildings are thus subjected is a sudden movement of the earth on which they rest in a certain direction, and then a sudden reversal of that movement. Mr. R. Mallet, the author of a well-known work on this subject, giving an elaborate account of the Neapolitan earthquake of 1857, found that the maximum amplitude of the wave, or amount of "shove" in a horizontal direction, did not exceed three or four inches.

The observed effects, as is well known, vary from a mere tremor to such a violent disturbance as to result in the overthrow of whole cities; and, as may be readily imagined, even in comparatively slight shocks the delicate apparatus in a lighthouse is peculiarly liable to derangement, and instances are recorded in which the lamp glasses have been shaken off and the lamps extinguished.

In 1867, the Board of Trade asked Messrs. Stevenson to advise as to establishing a lighthouse system on the coasts of Japan, which being a country peculiarly subject to earthquakes, they requested them specially to consider in what way the anticipated injury to the apparatus could be mitigated. To meet the difficulty thus raised, Mr. David Stevenson devised an Aseismatic-joint in the table supporting the apparatus and

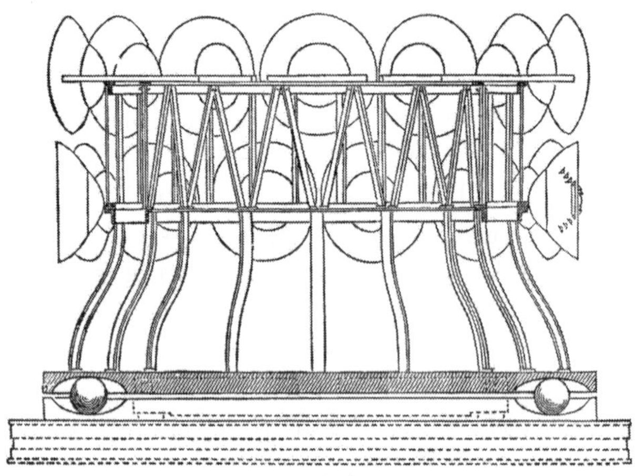

Fig. 135.

machine, so contrived as to admit of a certain amount of horizontal movement taking place in the building without affecting the table on which the apparatus rests, the idea being that in some horizontal plane the connection between the apparatus and the building should be cut through and separated, so that the sudden motion of the lower part should not be directly communicated to the superincumbent portion. Fig. 135 is a section in which the upper table bearing the holophotal reflectors is shown in hatched lines. This table rests on spherical bell metal balls, contained in cups formed in the upper and lower tables, and Fig. 136 is a plan of the lower table, showing the three points of support. The same

Fig. 136.

arrangement, on a larger scale, was proposed for the foundations of the towers, where these were intended to be constructed of iron. The effect of the aseismatic arrangement, which adds about £90 to the cost of the apparatus, was tested in the manufacturer's shop, by placing one of the tables on a movable truck, and driving it against buffers by the impact of heavy weights allowed to swing against it. The effect produced on a lighthouse lamp placed on the aseismatically arranged table was imperceptible, but when the aseismatic arrangement was clamped, the glass chimney was thrown off and the apparatus disarranged. The experimental trials having thus proved entirely satisfactory, several aseismatic tables were constructed and sent out to Japan. Those in charge of the apparatus however, complained of its sensitiveness, causing them inconvenience when engaged in cleaning or trimming the lamps. But this, as was pointed out in a report to the Japanese Government, could be easily cured by locking the tables while the lamps were being cleaned or trimmed.

Focusing of Lights.

Parabolic Reflector.—It was found by experiment that parabolic reflectors might be dipped to about $2\frac{1}{2}$ degrees below the horizontal without diminishing the power of the light sent to the horizon. In practice, however, it appears that no advantage is gained by dipping them more than about 1 degree. This amount, indeed, sends the central rays to within three-quarters of a mile of the foot of the tower when its height is 100 feet. All paraffin Argand burners are placed 14^{mm} below the focus, which brings the brightest part of the flame into the focus.

In Scotland the mode of testing the accuracy of the position of the burner is by an instrument having three arms crossing each other at the centre; the extremities of these arms are points in the paraboloid. Another arm, which represents the position of the axis of the parabola, betwixt the apex and the focus, passes through the point of intersection of the three arms. A small centre punch-mark is made on the paraboloid at its apex, and when the pointed end of the central arm is pushed in so as to touch this point, the other end is at the focus. A light plummet hangs from the focus end of this rod, which passes down through the burner, and on the thread is fixed a round ivory plug, on the surface of which, at the proper distance under the focus, a circle is engraved. The top of the burner is set exactly at the level of this line, the plug itself hanging at the same time exactly in the centre of the inner tube. Where the reflectors have been holophotalised the same instrument applies, the axial bar only being lengthened to go back to the centre of the spherical mirror.

Dioptric Lights.—The brightest horizontal sections of the flames of the different orders of lamps have been found by short expôsed photographs to be as follows :—

One-wick lamp 14 ^{mm} above the top of burner.		
Two-wick lamp 19 ^{mm}	do.	do.
Three-wick lamp 23 ^{mm}	do.	do.
Four-wick lamp 25 ^{mm}	do.	do.

These points are placed in the sea horizon focus of the central or refracting portions of the apparatus. The upper prisms are ground and set so as to bring the sea horizon to a focus for four-wick burners at a point 30 ^{mm} above the burner, and 9 ^{mm} *behind* the axis, and the lower prisms to a point 18 ^{mm} above the burner, and 38 ^{mm} *before* the axis ; in this way the brightest sections of the flame are sent to the horizon, and the bulk of the light is spread over the sea between the horizon and the lighthouse. These figures for three-wick burners are 29, 12, 17, and 23 respectively.

Mr. Alan Stevenson suggested the dipping of dioptric lights below the normal to a plumb line, in his Report of December 10th, 1839, to the Commissioners of Northern Lighthouses, in the following passage :—

" A more serious inconvenience in using catadioptric zones is, that in very high towers, where some correction of the position of the apparatus becomes necessary, so as to direct the rays to the horizon, the means of regulating the zones in a manner similar to that used for the mirrors is inapplicable.

" The adoption of a high point in the flame for the focus

of these zones, however, affords a considerable compensation for this defect; and it might even be entirely obviated by constructing each set of zones of the form suited to the known height of each tower and the required range of each light, if such a correction were found to be of sufficient importance to warrant its application."

But though the precaution of dipping the strongest of the light to the sea horizon was followed out by Mr. Alan Stevenson in high towers, it was not always attended to till the year 1860, when Mr. J. F. Campbell, the Secretary of the Royal Commission on Lighthouses, independently brought the subject prominently forward, and suggested the internal mode of adjustment. Since then the strongest beam has been invariably dipped to the horizon.

It must, however, be remembered that when the weather becomes even in the least degree thick or hazy, not to say foggy, the range of the light is greatly curtailed by atmospheric absorption and refraction; which last produces, during fogs, irregular diffusion of the light in every plane, so that at high towers, where the beam is pointed to a very distant horizon, it is obvious that the strongest light is directed to a part of the sea where it cannot be seen with certainty unless when the weather is exceptionally clear. It has lately appeared to me that the strongest beam should be dipped lower than Mr. Stevenson proposed, and as is now everywhere adopted.

The best of the light should certainly be directed to the place where the safety of shipping most requires it. Now, it may in most cases be laid down as axiomatic, that the peril of any vessel is inversely proportional to her distance from the danger, whether that danger be a lee-shore or an insulated rock. Confining ourselves to this one view of the subject, it would follow that the strongest of the light should, to suit hazy states of the atmosphere, be thrown as near to the shore or the rock as would admit of vessels keeping clear of the danger. But such a restriction as this would, if permanent, greatly impair the usefulness of the light by unduly curtailing its range in clear states of the atmosphere; and of course, *cœteris paribus*, the farther off a sailor is warned of his approach to the shore the better and safer. Besides, the loss due to atmospheric absorption increases in a geometric ratio, and as the rays diverge in cones from the apparatus the power of the light is further decreased in the inverse ratio of the squares of the distances from the shore. It is of course well known that the sun itself is extinguished by fog, and we cannot expect to compete with that luminary. But seeing there are endless variations in the density of fogs, and in the transparency of the air when there is no fog properly so called, it always appeared to me that, had we an easy way of doing so, we ought to increase *temporarily* the dip of the light, and thus, during haze and fogs, to direct the strongest beam to a point much nearer the shore than the sea horizon. At present we direct our strongest light, not only in clear weather, when it *can* be seen, but also during fogs, when it cannot possibly be seen, to a *part of the sea where the danger to shipping is in most situations the smallest;* and this is done to the detriment of that region where, even when the weather is hazy, there is at least some chance of the light being visible, and to a part of the sea where the danger to shipping is unquestionably the greatest.

The simplest mode of depressing the light temporarily would be to raise the lamp itself in relation to the focal plane of the lens. But this is, for several reasons, very inexpedient. The proper adjustment of the apparatus to the focus, so as to secure its being situate in the section of maximum luminosity of the flame, is a somewhat delicate one, and ought, if possible, not to be disturbed oftener than is necessary for changing the lamp. Moreover, while the raising of the lamp would depress the light which passes through the refracting portion of the apparatus, it would have precisely the opposite effect upon the portions which pass through the totally reflecting prisms placed above and below the refracting part, which would then throw the rays upwards to the sky, where they would be useless. But any desired change could be effected by surrounding the flame with prisms spheric on their inner faces, and concentric with the foci of the different parts of the apparatus, so as to depress the rays before they fall upon the main apparatus. Those prisms which subtend the lens would have their thicker ends lowest, and those subtending the reflectors would have their thicker ends uppermost.

From a series of observations made with two kinds of photometer by Messrs. Stevenson, in 1865, on the penetrative power of light from a first-order lens and cylindric refractor, it appears that for an angle of 0° 30′ in altitude above the plane of maximum intensity, and for 0° 30′ below that plane, the power of the light does not vary more than at greatest from 4 to 6 per cent, and that, if the strongest part be sent to the horizon, about one half of the whole is sent uselessly to the skies.

Power of Lens in the Vertical Plane.

			Mean of four sets of observations.
0° 40′	ABOVE the level of	maximum	·90
0 30′	do.	do.	·94
0 20′	do.	do.	·97
0 10′	do.	do.	·98
0 0′	maximum power	.	1·00

			Mean of four sets of observations.
0° 10′ BELOW the level of maximum			·99
0 20′	do.	do.	·97
0 30′	do.	do.	·96

Note.—These results, which are the mean of four sets of observations did not extend further in the vertical plane.

Result of Dipping Light as proposed, contrasted with present System.

PRESENT SYSTEM.		PROPOSED SYSTEM.	
Above horizon.	Power.	Above horizon.	Power.
0° 40′	·90 lost on sky.	0° 20′	·90 lost on sky
0 30′	·94 do.	0 10′	·94 do.
0 20′	·97 do.	0 0′	·97 on horizon.
0 10′	·98 do.		
0 0′	1·00 on horizon.	Below horizon.	
		0° 10′	·98 on sea.
Below horizon.		0 20′	1·00 do.
0° 10′	·99 on sea.	0 30′	·99 do.
0 20′	·97 do.	0 40′	·97 do.
0 30′	·96 do.	0 50′	·96 do.

Applying these observations, so far as they extend in the vertical plane, to the case of lighthouses elevated much above sea-level, we see that to dip the strongest beam to a point much nearer the shore than the sea horizon, while it would not appreciably affect the visibility there, would even, so far as the observations go, increase the power of the light nearer the shore. Those who have been close to a lighthouse on a hazy night must have noticed the luminous rays passing through the air far above the sea-level, and cases are adduced by Mr. Beazeley of shipwrecks having occurred when the light could not be seen by the sailors, although their vessels were stranded close to the tower. As the lens has the greatest divergence, and is the only agent for giving light near the shore, it only should be dipped so as to throw as few of the rays as possible uselessly on the skies, while the reflecting prisms, which have much less divergence, will remain as at present throwing their rays to the horizon. By this different distribution of the light from the lens and the prisms, by which the strongest beams from the lens are dipped 0° 20′ below the horizon, which causes a loss there of 3 per cent of lens power, yet the loss on the *whole* light coming from both lens and prisms, taken at 70 and 30 respectively, will be reduced to only about 2 per cent, while the sea near the shore will be more powerfully illuminated than at present. It may, however, be fairly questioned whether the strongest beam ought not to be dipped to 0° 30′, as this would still further increase the power near the shore, and would only depreciate the light at the horizon by about 5·8 per cent. It is well to remember that, should the flame, through the neglect of the keeper, fall at any time below the standard height, such a defect would operate most injuriously on the light falling near the shore, and not so much on that sent to the horizon. Now there can be no question that in all ordinary cases a vessel with such an offing as 20 miles, which is the sea-range due to 300 feet of elevation, is in a far safer position than if she were within a mile or two of the shore, and hence the propriety of increasing the light near the shore, so long as we do not to any appreciable extent reduce it at the horizon.

Optical Protractors.

Mr. Balfour's Protractor.—The late Mr. J. M. Balfour, in the description of his protractor, which was the first of the kind (vol. v. Trans. Royal Scot. Soc. Arts), refers to a method described in Dr. Young's Lectures on Natural Philosophy, of finding the path of a refracted ray, and on that principle he constructed an instrument for tracing the form of totally reflecting prisms. Dr. Young's method is as follows :—

A B C (Fig. 137) is a circle touching the refracting surface at A, if another circle D E F be drawn on the same centre, having its diameter to that of the first as the sine of the angle of incidence is to that of refraction, and a third circle, G H I, which is less than the first in the same proportion as the second is greater; and if the direction of the incident ray, K A, be continued to D, and O D be drawn from the common centre O, cutting G H I in G, A G will be the direction of the refracted ray.

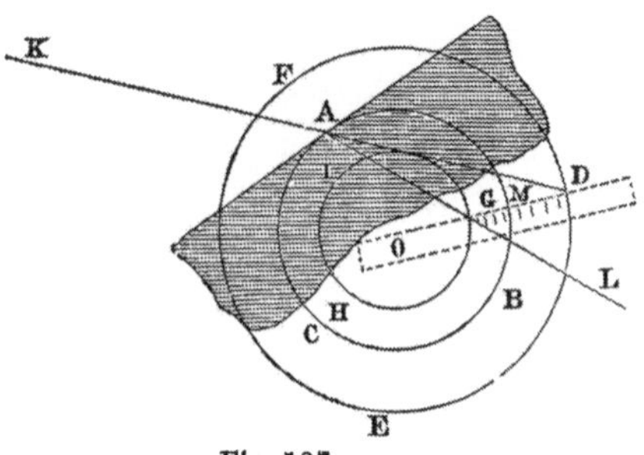

Fig. 137.

In the figure O D is a ruler, turning on a centre O. M, the point where the ruler cuts the circle A B C, is marked 1·0. D is at 1·5, the index of refraction for lighthouse glass.

To use the instrument, draw a perpendicular to the surface of the medium from the point of incidence A, insert a needle at the point O in this line, making the distance OA = OM. Turn round the ruler till the division on the outer scale corresponding to the index of refraction (1·5 in the figure) cuts the incident ray produced, and make a mark at the corresponding division on the inner scale. A line drawn from A through this mark shows the direction of the ray through the medium.

The principle of this protractor is applied to the construction of totally reflecting prisms in the following manner:—

M N (Fig. 138) is the ruler marked, as already described, for an index of refraction of 1·5. F C is the direction of a ray from the flame which has to be bent into the direction C E. The sides of the prism C Q and C P must have such directions that the ray F C falling on the side C P at the point C will be refracted into the direction C Q, and C P must be such that the ray E C traced backward must, after falling on C Q, be refracted in the direction C P. This is found by the instrument shown. The dotted line through C L bisects the angle E C F, and the point L is moved to or from the point C on the line C L, until it is observed that the end of the ruler M N just touches the line F C produced, while the mark 1·5 coincides with the line C Q. The lines C Q and C P then form the two refracting sides of the prism; the reflecting side is drawn, so that all the rays falling upon it are reflected in directions parallel to the line C P.

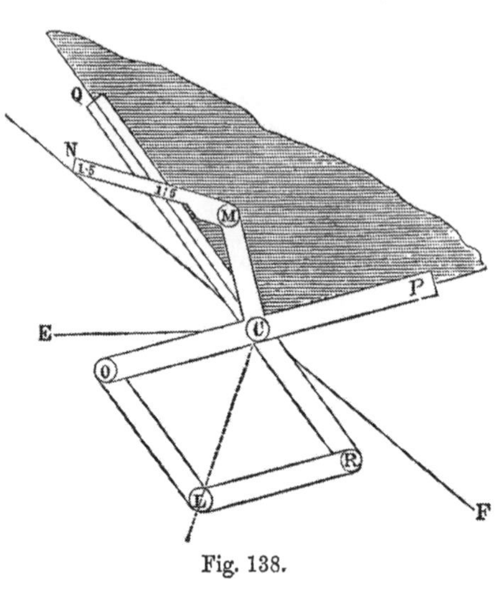

Fig. 138.

Mr. Brebner's Optical Protractor.—This instrument, as shown on Plate XXV., consists of two parts: the lower limb L H N G, which forms a complete circle, and the upper limb,

A K B, which is a semicircle. The lower limb is made to move easily round a centre on the upper limb, which is perforated and filled in with glass, and two axial cross lines are drawn on the glass. The under side of this centre part rests on the table, and is fitted with three points which penetrate the table and keep the upper limb from turning; while the lower one, which is next the table, turns freely. One half of the lower limb is divided into degrees, and is numbered from the middle towards each side, and also from each side towards the middle; the opposite half is so divided as to make the sines of the opposing angles have a ratio of 1·51 to 1.

For finding the path of a ray through a piece of glass: Let G H represent the surface of a piece of glass, and F C (F is placed out of position in the plate) any ray falling upon it. Read off the angle from the perpendicular L C N, and mark off the same reading from N towards I. In Plate XXV. the ray F C, which cuts the limb at 19½°, will pass through the glass in the direction C P, which is 19½° from N.

For constructing a Totally-reflecting Prism.—Suppose it is required to construct a prism which shall reflect into a parallel beam, in a required direction, the angle of light C F M, emanating from a flame F. The point F, the apex of the prism C, and the required direction C T being fixed, place the centre of the instrument on C, with the axis of the upper limb A B coinciding with the line of the emergent ray T C, and keep the upper limb in this position by slight pressure, while the instrument is being used. Now move round the lower limb, bringing the perpendicular C L towards the ray F C, until the angle between L and the ray F C and its complement (which to save performing the deduction mentally is marked above it) reads the same as the angle from A towards

K, and the angle from N towards S. For example, in the figure the ray F C which cuts at 19½°, the complement of which is 70½°, and 19½° from N towards S coincides with 70½° from A towards K, and in no other position of the two limbs can this occur. G H and C P then fix the two sides of the prism, and the angle T C P will be equal to the angle F C H. To determine the reflecting side, draw the dotted line E C parallel to F M, observe where it cuts the lower limb at X, viz. 10¾°, and mark off the same reading at Y. Bisect the angle Y C H, and from M draw the line M W perpendicular to the bisecting line, until it meets the medial bisecting line C R; then bisect the angle P C H, and from the point W draw the line p W perpendicular to the bisecting line; p W and W M are the tangents to the curve p O' M, and all the rays falling upon the side C M from the focus F will be refracted so as to fall on the side M O' p, at such an angle as to ensure their being reflected in a direction parallel to M C, and will consequently be refracted at the side C p, so as to emerge from it parallel to C T and p O.

PHOTOMETERS.

Many photometers have been proposed, the simplest of which is the comparison of the shadows of two lights which are moved until the shadows are equal. The relative intensities are of course ascertained by the law of inverse squares of the distance.

In Plate XXIV. are shown two forms, which I proposed in 1860. Fig. 3 shows, in cross section and elevation, a liquid photometer, in which A B C is a compressible india-rubber bag, filled with diluted ink, and in order to alter the length of fluid passed through by the light there is a milled headed screw, E, for compressing the bag, in which there are fitted two pieces of glass. Fig. 4 shows a photometer on a different principle, which I described in 1860, in which rays of light are admitted through a small hole *a*, and allowed to diverge into the tube *c b d e*, the distance through which the light diverges is measured by the distance to which the sliding tube *h g f h i l* has been drawn out. By this mode, as the light is made divergent, its intensity is reduced in the inverse ratio of the squares of the distances. Fig. 5 shows another instrument on the same principle, in which a lens is substituted for the small hole in the diaphragm of the photometer last described. The latest arrangement which I have adopted is that shown in Fig. 139. It is a box with openings in two adjacent sides, through which the rays pass from the illuminant L, used for the purpose of comparison, and from L′ the light to be compared. The rays, after passing through these openings, fall on two planes at the points *d* and *e*, which are directly beneath the eyepiece *f*, through which the operator makes his observation. The planes on which the rays fall are inclined at 45° to the horizontal, and 90° to the vertical planes of the rays. The opening through which the ray from L passes is of constant area, and the illuminant should be placed at a fixed distance from that opening. The opening *b* through which the ray from L′ passes can be enlarged or contracted by means of two opaque plates moving

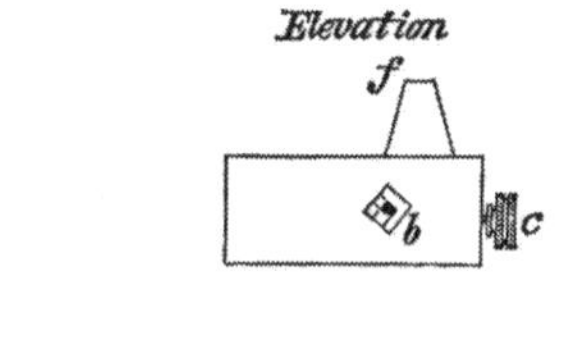

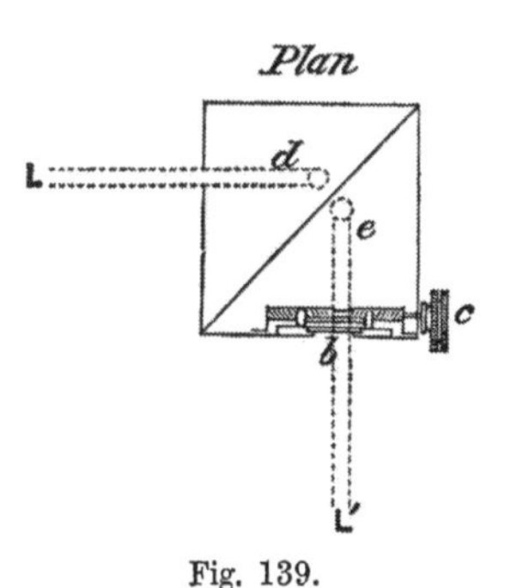

Fig. 139.

diagonally and attached to the double screw *c*. On one of the plates there is a scale which shows the proportion of the light admitted. When it is wanted to find the power of a light, a candle, Carcel burner, or other known unit of light is placed at a fixed distance from the fixed opening; the variable opening *b* is then brought round till the light to be judged of shines directly through it, the screw *c* is then turned till both lights appear of equal intensity, when the number on the scale gives the power of the light.

Time of Lighting and Extinguishing Lighthouses.

The lamps are lit in most lighthouses at sunset and extinguished at sunrise, but in northern latitudes, where there are long periods of twilight, this is quite unnecessary, and would occasion a needless expenditure of oil. Mr. Robert Stevenson, with the assistance of the late distinguished Professor Henderson, Astronomer Royal of Scotland, determined the mean duration of the twilight in different latitudes, from which tables were calculated for every day in the year. The Scotch lighthouses are therefore lighted from the going away of daylight in the evening till its return in the morning, unless the weather be dull, when the keepers exhibit the lights at an earlier period. Tables of the times of lighting and extinguishing for each day of the year are hung in each lightroom. By the adoption of this plan of lighting an annual saving resulted for the Scotch lights, in 1861, of £1600, representing, at 5 per cent, a capital sum of £32,000 —(Royal Commission Report, 1861, vol. i. p. 189.)

CHAPTER X.

FOG-SIGNALS.

FOG-SIGNALS are of two kinds, (1) the *luminous*, which are produced during fogs by accessions to the light which is usually shown in clear weather; and (2) those which are produced by *sound*.

LUMINOUS FOG-SIGNALS.

Different modes have been proposed for temporarily increasing the power of a light when fog comes on. So far as I know, the first of the kind was by Mr. Wilson of Troon, in Ayrshire, in 1826. The effect was produced by the simple method of increasing the supply of gas and the number of burners. Sir David Brewster, in 1827, proposed special optical apparatus which could occasionally be brought into action when the state of the atmosphere required it. The exhibition of Bengal and other lights was also proposed by him and others. Mr. Wigham has very successfully introduced at several places in Ireland fog-signals produced by separate optical apparatus, placed one above another, each of which is lighted by a burner of large size. The excellent effect produced has been proved on the authority of Professor Tyndall, as well as by captains of vessels which regularly pass along the coasts where these lights have been established. Professor Tyndall has found that when other lights were invisible the illumination of the fog was not only sufficient to attract the attention of the sailor, but that even the variations in the volume of the flame were distinctly visible in the haze.

Fig. 140.

Fig. 140 represents that form of Mr. Wigham's apparatus which he terms the quadriform arrangement, in which four separate burners and lenses are employed. The burners have been already described in Chapter VIII.

The optical objection to such arrangements is sufficiently apparent, for the only rays from each burner that are correctly utilised are those which fall on the lens which has its focus in that burner. This objection must, however, be regarded as of less importance than the advantage of having, at immediate disposal, when required, so large an amount of light. In every case the practical result must certainly be looked on as more important than the merely economic. It occurred to me in 1875 to design apparatus which could be applied to more than one flame, so as to carry out Mr. Wigham's proposal without any loss of light, and although such an arrangement may be not very suitable to an ordinary flame, it would be quite applicable to the electric light. In Fig. 141 ff' are two flames, L the upper and L′ the lower lenses or cylindric refractors. The rays escaping below the lenses L′ of the lower apparatus are returned again to the focus f' by pieces of spherical mirror *m*, while at the top of the lower lenses or refractors there are placed "Back Prisms" P (Chap. III.) These

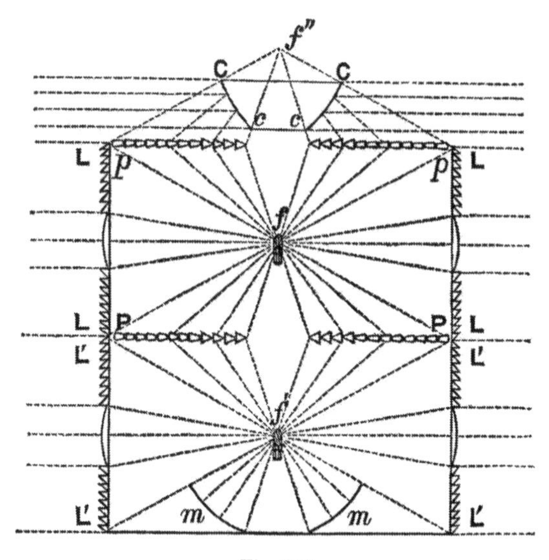

Fig. 141.

prisms will, in this case, have conjugate foci situate in the upper and lower flames (f and f'), so that the light escaping above the lenses or refractors of the lower flame f' and below those of the upper flame f, will be intercepted and utilised. The rays escaping above the lenses or refractors of the upper flame f will be intercepted by similar prisms *p p*, the conjugate foci of which will be situate respectively in the upper flame, and in a point f'' equidistant above the prisms P. The rays converging towards this upper focus f'' will be intercepted by a parabolic conoid C *c*, of novel form, presenting its *convex*, and not as hitherto its *concave*, surface to the rays. This compound arrangement for the upper rays might be avoided by the ordinary cupola of reflecting prisms; but what has been described, though less perfect, is more compact. The optical property of this instrument is to render parallel the converging rays of light which fall upon it. In this way the separate flames and agents will be optically combined, so as to utilise all the light.

Mr. Douglass's burner, of single and double power, for use in thick weather, has already been described in Chap. VIII.

Mr. Douglass's Movable Lantern.—The prevalence of fogs at high levels, when places at lower levels are free from it, is well known, and in order to meet this evil Mr. Douglass introduced at the South Stack Lighthouse, Island of Anglesea, a portable lantern, which moves down a railway, so as, during fogs, to give its light close to the level of high water. The arrangement is shown on Fig. 1, Plate XXXII.

SOUND FOG-SIGNALS.

The various modes of producing sound for fog-signals are very briefly given in the following statements, many of which are taken from Mr. Beazeley's paper in the Transactions of the Institution of Civil Engineers:—

Bells.—In 1811, two bells were supported on standards on the balcony of the Bell Rock Lighthouse, for the purpose of diffusing sound all round. They are tolled by the same machine which causes the lighting apparatus to revolve. Similar bells were used at the South Rock Island in 1812, and since that date they have been applied at many rock stations.

In 1811, Mr. R. Stevenson designed a beacon for the Carr

Rock, already referred to, which was surmounted by a bell rung by a train of wheels, and actuated not by the waves, but by the rising and falling tide. The tide water was admitted through a small aperture, 3 inches in diameter, perforated in the solid masonry at the level of low water, and communicating with a close cylindrical chamber in the centre of the building, in which a float was to rise and fall with the tide. A working model on this principle was constructed and kept in motion for several months, but unfortunately, as already described, on account of its exposed situation the beacon itself was never completed. In 1817, when it was resolved to substitute an iron for a stone beacon, Mr. R. Stevenson found that he could not apply a bell, and tried some experiments with a whistle blown by air compressed by the rise and fall of the tide; but he did not consider the result sufficiently satisfactory to warrant its adoption. He also thought of employing a phosphoric light, but no trial of it seems to have been made.

For the purpose of distinction, Mr. Douglass, at the Wolf, in 1870, employed a bell struck three times in quick succession every 15 seconds. At Dhu Heartach, in 1872, Messrs. Stevenson introduced another distinction, viz. a bell rapidly rung by rotating hammers, producing a *continuous rattle* for 10 seconds, the intervals of silence being 30 seconds; and in 1875 Messrs. Stevenson also employed at Kempock Point, on the Clyde, two bells of different notes, struck immediately one after the other, to distinguish Kempock from a neighbouring station where there was a single bell. These bells, it may be remarked in passing, are worked by gas-engines. Bells are also frequently fixed on buoys, which are intended to be rung by the motion of the waves; but owing to the necessarily heavy construction of these buoys they do not ring unless the sea is greatly agitated. To obviate this, I proposed, in 1873, to suspend under the water, and close below the buoy, a heavy sinker, which arrangement virtually forms a shorter pendulum, thus increasing the number of oscillations in a given time. Bells used as fog-signals vary from 3 cwt. to 2 tons in weight.

Gongs, struck by hand with padded hammers, are largely used, especially on lightships, but they produce very feeble sounds, and are rapidly being superseded by more powerful apparatus.

Guns, fired at intervals, are largely employed as fog-signals, but they are not very efficient, not only on account of the long interval which must necessarily intervene between the shots, but the report, although very effective to leeward, makes little or no headway against the wind. This is in accordance with the ascertained fact that the best carrying sounds are those which are prolonged or sustained for some seconds, such as are produced by whistles or sirens, and the report of a gun is of course defective in this respect, being short and sharp. The guns used for this purpose are 12 or 18 pounders, with 3 lbs. charges of powder, fired at intervals of from 15 to 20 minutes. Mr. Wigham has with considerable success used a charge of gas instead of powder.

Rockets.—Experiments were made in 1877 by the Trinity House on the use of explosive rockets as fog signals. These rockets (2 oz. to 8 oz.), which are made of either cotton powder or gun cotton, on reaching a certain elevation, explode with a loud report, and the results, as reported by the Elder Brethren and Professor Tyndall, were very encouraging, finding, as they did, that the rockets were better heard than the gun, though less effective than the siren. At rock Lighthouse stations the Trinity House thought they might be used with advantage in place of bells, provided the material could be safely stored and fired. The great space which is required for the ordinary class of fog-signals, besides other objections, precludes the possibility of employing them at rock stations; rockets, on the contrary, can be stored in any lighthouse tower.

Whistles.—The late Mr. A. Gordon in 1845 suggested the combination of a steam whistle with a reflector. He also proposed to place a whistle sounded by air or steam in a Bordier Marcet reflector, or in the focus common to three parabolic reflectors grouped together, and revolving on a vertical axis.

Mr. Daboll in 1850 devised a whistle sounded by compressed air, which was tried at Beaver Tail Point with excellent results.

In 1856, Mr. W. Smith patented a combination of four whistles, with their sounding edges straight in plan instead of circular.

In 1860, Mr. E. A. Cowper designed a fog-signal in which he proposed to employ a whistle 15 inches in diameter, placed in the focus of a parabolic trumpet 25 feet long. He also suggested the propriety of making the intervals and character of fog-signals to correspond with the character of the light to which it is auxiliary.

In 1860, Mr. Beazeley, in his paper (Min. Instit. Civ. Eng.) stated that whistles might produce a lower and higher note alternately. This proposal was carried out at the Cloch Lighthouse on the Clyde in 1874, on the recommendation of Messrs. Stevenson, and is believed to be the only signal of the kind which is now in use.

In 1861, Mr. Barker proposed to use superheated steam for sounding a fog-signal, which is now in operation at Partridge Island, New Brunswick, and other places.

Whistles, varying in diameter from 2½ to 18″ in diameter, were experimented on in America, with the result that those of 10″ to 12″ were recommended for ordinary use. The best results were obtained with whistles of the following proportions :—diameter of bell equal to ⅔ds of its length, and the vertical distance of the lower edge above the cup from ⅓ to ¼ of the diameter, for a pressure of from 50 to 60 lbs. of steam.

At Partridge Island, New Brunswick, the steam-whistle is between 5″ and 6″ diameter, with a pressure of 100 lbs. per square inch. At Aberdeen harbour the whistle is 6″ diameter, and the pressure 60 lbs. to the inch; and in Professor Tyndall's experiments, to be afterwards referred to, the largest whistle used was 10 inches diameter, 18 inches high, with a pressure of 70 lbs. to the inch.

There have been at different times various suggestions made for increasing the power of fog-whistles and for working the apparatus automatically, but the whistle has in a great measure been superseded by the more powerful siren and reed trumpet.

Courtenay's Automatic Whistling Buoys.—This excellent device has been already described in Chapter V.

Reed Trumpet.—In this instrument the sound is produced by compressed air escaping past the edges of a vibrating metal tongue or reed placed in the small end of a trumpet. In 1844 Admiral Tayler suggested reed pipes sounded by condensed air to produce the sound for fog-signals. Mr. Daboll, in 1851, designed a trumpet having a metallic reed, also sounded by compressed air, one of which was placed at Beaver Tail Point in America. It is said to have been distinctly heard at 6 miles distance against a light breeze, and during a dense fog; and Captain Walden states that he heard it at $2\frac{1}{2}$ miles distance, in a rough sea and with a strong cross wind. Mr. Daboll also patented, in 1860, the application of condensed air to the sounding of whistles or horns, with machinery to produce the revolution of the horn whilst sounding. The Daboll horn has been introduced at several lighthouse stations in America and in this country.

In 1863, Professor Holmes devised a trumpet, having a reed of German silver sounded by compressed air, with an automatic arrangement whereby the trumpet can be placed at a distance from the air compressor and engines. This trumpet is constructed to vibrate in unison with its reed, and it has been successfully established at several lighthouse stations.

In Professor Tyndall's experiments the reed was 11 inches long $\times\ 3\frac{1}{8}'' \times \frac{1}{4}''$, and at Pladda it is $1\frac{7}{8}$ inch broad, $8\frac{3}{8}$ inches long, and $\frac{1}{4}$ inch thick. Mr. Beazeley recommends for a high pressure reed $5\frac{1}{2}'' \times 1'' \times \frac{5}{16}''$, and for a low pressure reed $6\frac{1}{2}'' \times 2\frac{3}{4}'' \times \frac{5}{16}''$.

Siren.—This instrument is sounded by steam or compressed air, escaping in a series of puffs through small slits in a disc or cylinder caused to revolve with great rapidity, and fixed in the small end of a trumpet.

In 1867, Mr. Brown of New York made experiments with the siren, and Professor Tyndall subsequently experimented with this instrument; and in 1874 he reported to the Trinity House that the "siren is, beyond question, the most powerful fog-signal which has hitherto been tried in England. It is specially powerful when local noises, such as those of wind, rigging, breaking-waves, shore-surf, and the rattle of pebbles, have to be overcome. Its density, quality, pitch, and penetration, render it dominant over such noises after all other signal sounds have succumbed." Professor Tyndall, in his report, already referred to, states that it "may with certainty be affirmed that in almost all cases the siren, even on steamers with paddles going, may be relied on at a distance of 2 miles. In the great majority of cases it may be relied upon at a distance of 3 miles, and in the majority of cases at a distance greater than 3 miles."

In Tyndall's experiments the diameter of the disc was 4½ inches, making sometimes 2400 revolutions per minute; and at St. Abb's Head and Sanda the diameter is 12½ inches, making 1464 revolutions per minute.

The trumpets, which have been referred to as being used with reeds and sirens, vary from about 8 feet long, 2 feet diameter at mouth, and 2 inches at small end, to about 17 feet long, 3 feet diameter at mouth, and 3 inches at small end. The pressures with which they are sounded vary as much as from 5 to 50 lbs. per square inch; about 20 lbs. is, however, a very effective pressure.

Motors.—For the purposes of compressing the air for reed trumpets or sirens, and for actuating the hammers of bells, etc., various motive powers have been proposed or adopted, of which the following is a summary:—

(1.) Clockwork driven by the descent of weights or the action of springs.

(2.) Water-power applied to wheels, turbines, and water-engines of various kinds.

(3.) Rise and fall of the tide; advantage being taken of the hydrostatic pressure thereby produced, or of the natural tidal currents, or of artificial currents produced by impounding tide water; or lastly, of the motion of waves.

(4.) Steam-engines of various types.

(5.) Hot air engines, either as designed by Professor Holmes on Stirling and Ericsson's principle, or as constructed by Messrs. Brown of New York, in which the *working air* is heated not by mere contact with a metallic surface, but by being itself passed through the fire (which is therefore under pressure), and supports combustion.

(6.) Gas engines worked by coal gas, Keith's mineral oil gas, and air gas.

The first of these being of small power is principally used for ringing bells; the second and third are applicable only in situations where the necessary physical characteristics present themselves; and the fourth requires a large volume of water, not always to be got at lighthouse stations, is liable to explosion in unskilful hands, and takes a long time to start, which is a very objectionable feature, as fogs frequently come on suddenly. This last objection also applies to hot air engines, which require no water, and are otherwise unobjectionable; but, while the gas engine is free from this objection, it, on the other hand, requires a small supply of water. It will thus be seen that it is impossible to generalise as to which form of motor is the best, but each case must be considered on its own merits. The horse-power necessary at a fog-signal station depends on the pressure, duration, and frequency with which the blast is emitted, as well as the range or distance at which it is desirable that the sound should be heard. It varies from 8 to 20 horse-power.

The investigations of Professors Henry, Tyndall, and Stokes, have elicited very important facts, which show that, independent of the efficiency of the generating apparatus, there are disturbing causes in the atmosphere itself, which at certain times modify the effects of fog-signals of every kind.

Although the findings of Henry and Tyndall may be regarded as to a large extent the same, the causes which influence the unequal transmission of sound are differently accounted for by these philosophers. Professor Henry explains the effects as due to the co-existence of winds blowing in different directions in the upper and lower strata of the atmosphere; while Professor Tyndall, whose extremely interesting narrative of the investigations made by him for the Trinity House is well worthy of attentive perusal, attributes the unequal transmission of sound in different directions to what he has termed flocculence, or want of homogeneity in the density of the atmosphere.

Professor Stokes has proved the important fact that the inequality and the velocity of the wind as we ascend into the atmosphere have necessarily the effect of disturbing the regularity of the transmission in spherical shells. He states that "near the earth, in a direction contrary to the wind, the sound continually tends to be propagated upwards, and consequently there is a continual tendency for an observer in that direction to be left in a sort of sound shadow. Hence, at a sufficient distance, the sound ought to be very much enfeebled; but near the source of disturbance this cause has not yet had time to operate, and therefore the wind produces no sensible effect except what arises from the augmentation on the radius of the sound-wave, and this is too small to be perceptible." This result should be kept in view in dealing with places at high levels above the sea.

As these atmospheric phenomena cannot of course be overcome, being entirely external to the apparatus which produces the sound, it will not be necessary to occupy space in this manual with details of the results, which are, however, of a very interesting character.

The results arrived at by Professor Henry will be found in the Reports of the American Lighthouse Board and the Smithsonian Institution. The important narrative of the investigations undertaken by Professor Tyndall in 1873 for the Trinity House is given in his Reports to that Board.

Professor Henry, in his Report of 1874, speaks somewhat disparagingly of reflectors for condensing sound-waves, though he states that better results might be expected were the reflector made to intercept more of the waves of sound than the reflector with which he made his experiments. There can be no doubt, however, that if such instruments could be made of sufficient size to suit the amplitude of the waves of sound, and rigid enough to reflect and not absorb them, they would enormously increase the intensity in the direction of their axes. The experiments about to be described, with a small sized reflector at Inchkeith, showed a greatly increased effectiveness at moderate distances from the instrument, and there seems every reason to expect that the intensity of sound would increase with the dimensions of the reflector.

In 1866 I suggested the plan of a reflector for utilising all the sound waves (as in the case of light for illuminating apparatus for lighthouses), under the name of the *Holophone*, which is thus described in Roy. Scot. Soc. Arts. Trans. (12th February 1866) :—

"Fig. 142 represents the front elevation of an arrangement for diffusing the sound over 180°, and Fig. 144 that for condensing the sound into one beam. The action of these instruments will be easily

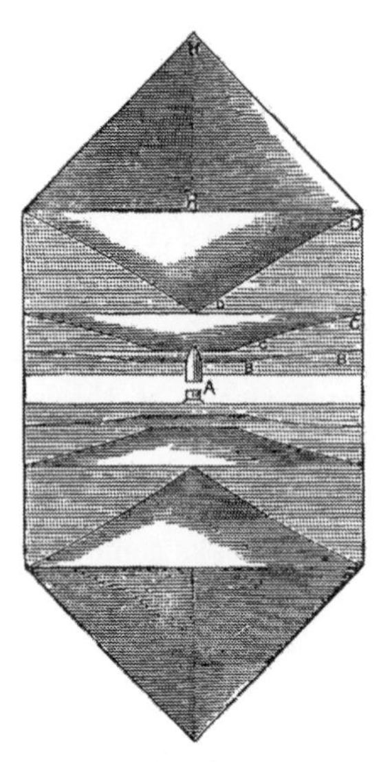

Fig. 142.

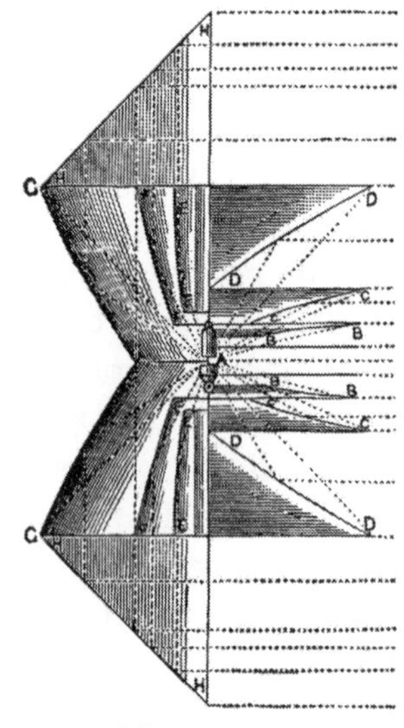

Fig. 143.

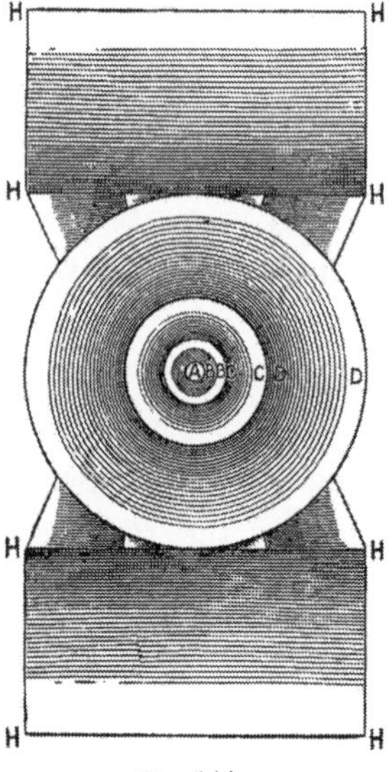

Fig. 144.

understood on looking at the diagrams. Figs. 142 and 143 are elevation and section of reflectors, which parallelise the rays in the vertical plane only 'B C D are paraboloidal strips for acting on the waves of sound produced by the anterior half of the whistle, or other radiant placed in the focus. E F G are similar strips for parallelising the waves proceeding from the posterior half of the radiant. H H are hollow cones for reflecting forward the rays parallelised by E F G.

"Fig. 144 represents the holophone which reflects the sound in every plane, and H H are the side reflectors, which in this case, instead of being conical are plane, and placed at an angle of 45° with the lateral beams of sound.

A is the whistle, bell, siren, or whatever radiant is to be employed. The whistle may either be placed vertically, as shown in the drawings, or horizontally, should that position be found preferable. In order to reduce the loss of sound passing off by the vibrations of the metal itself, the backs of the reflectors should probably be lined with plaster of Paris or some other suitable material. To render the reflector practically useful, it should of course be mounted on an axis (as has been done in the large horn recently constructed for the Trinity House of London by Mr. Daboll), so as to admit of being turned round, and thus, like a revolving light, to reach successively every azimuth. It will be observed that in these designs a hemispherical reflector has not been adopted for returning the vibrations through the focus, as they might to a great extent be destroyed by impinging on the whistle, or other instrument placed there. By the arrangement shown the sound waves are reflected in such a manner as to meet with no interruption.

"By directions of the Commissioners of Northern Lighthouses, and with the approbation of the Board of Trade, a holophone 8 feet by 16 feet, was constructed of malleable iron about one-sixteenth of an inch thick.

"The result of experiments tried at Inchkeith, which were made with a *low* power of steam was, that, while at moderate distances of about 1½ mile, the holophone was very much louder than a naked whistle supplied from the same boiler, the difference between the maximum distances at which each could be heard was less than might have been expected from the results obtained at lesser distances. Thus, while at the distance of 2 miles the holophone was very much louder, the naked whistle died out at 3 miles, and the holophone between 3 and 4 miles. It must be kept in view that the great use of fog-signals is for comparatively short distances, and for that the holophone was far more efficient than the simple whistle."

Fogs in this country average about 6 hours in duration, but often last 20 hours, and the longest recorded in the Northern Lighthouse service is 70 hours; while the total number of hours of fog per annum is about 400.

MEANS OF DISTINGUISHING VESSEL'S FOG-SIGNALS FROM THOSE AT LAND STATIONS.

It seems highly desirable that there should be a radical distinction between the fog-signals used by vessels to warn them of their approach to each other, and those signals employed to mark headlands which vessels are to avoid; and so long as no law exists for the regulation of the signals which are to be employed on board of ships, it is not possible to insure such a distinction as to make their sound clearly distinguishable in all cases from a fog-signal at a land station.

Messrs. Stevenson, in a Report to the Northern Lights Board, suggested that this difficulty might be overcome were some legislative enactment passed that the signals made by steamers afloat should in all cases be steam whistles of a shrill character, like that of a locomotive, while those used at lighthouse stations should be, as at present, the deeper sound of the fog-horn or siren; or if whistles are employed they should give two different musical sounds, as at the Cloch Lighthouse on the Clyde.

MR. A. CUNINGHAM'S OBSERVATIONS OF THE RELATIVE VISIBILITY OF DIFFERENT COLOURS IN A FOG.

Mr. Cuningham, late Secretary to the Northern Lighthouse Board, instituted a series of experiments on this subject at 11 of the Scotch Lighthouse stations, by means of a horizontally placed spar of wood, coloured black, white, and red, and the distances at which the colours ceased to be visible were observed by the lightkeepers, with the following results :—

	Black.	Red.	White.
St. Abb's Head	178	135	110
Kinnaird Head	239	223	154
North Ronaldsay	361	361	335
Sumburgh Head	202	149	174
Whalsey	547	501	443
Dunnet Head	159	165	135
Cape Wrath	395	356	307
Skerryvore	575	487	390
Mull of Kintyre	239	227	227
Pladda	486	441	324
Calf of Man	330	281	262
	3711	3326	2861
Mean	337	302	260
Ratios	1·00	0·90	0·77

It may be deduced from these observations that, when practicable, without interfering with any uniform system, those beacons and buoys which occupy the most important positions should be either black or red.

APPENDIX.

MATHEMATICAL FORMULÆ.

THE object which I kept in view in the preceding pages was to explain the general principles on which every lighthouse apparatus is designed—the optical agents which have from time to time been invented, and the combination of those agents, so as to form arrangements by which the optical problems of the lighthouse engineer have been solved. It would be obviously out of place, in a practical handbook such as this, to give the different steps in the mathematical theory of those optical instruments. All that appears necessary for the engineer is to be put into possession of the more important formulæ, which are accordingly appended. For those who wish to satisfy themselves as to the steps which lead to these formulæ, reference must be made to some of the following authorities.

Augustin Fresnel, in addition to his invaluable inventions, which have already been fully described in Chapter II., investigated mathematically the principles on which they depend.

Alan Stevenson, besides many improvements which he originated, and to which reference has already been made, gave, in the "Notes on Lighthouse Illumination," which accompany his "Account of the Skerryvore Lighthouse" in 1848, the first published account of the formulæ of Fresnel, to which he added mathematical investigation and tables of the calculated elements of totally reflecting prisms for lights of the first order. In his "Rudimentary Treatise on Lighthouses," published by Weale in 1850, these were again published.

Professor Swan, in his memoir "On New Forms of Lighthouse Apparatus," in the Transactions of the Royal Scottish Society of

Arts 1868, besides his independent invention of the back prisms (Chap. II.), and his proposal of the method of sending through the interspaces of apparatus, light coming from other apparatus placed behind (Chap. II.), has also devised several other dioptric lighthouse agents (Chap. II.)

He has, moreover, given a very complete mathematical analysis of the formation of a great variety of lighthouse prisms, including in the more important cases the consideration of excentric rays, so that no light may escape reflection. To him also is due the first mathematical investigation of the double reflecting prisms employed in the dioptric spherical mirror, as well as of the "back prisms," (Chap. II.) Looking to the extent and character of his work, he may be regarded as having contributed more largely than any other writer since Fresnel to the mathematics of lighthouse illumination.

Mr. J. T. Chance published other modes of investigations relating to the instruments which were then known, in the Minutes of Proceedings of the Institution of Civil Engineers in 1867, (vol. xxvi.)

M. Allard's formulæ for lighthouse statistics have already been given in Chapter VII. Besides these, reference may be made to *Reynaud* "Sur l'eclairage et les Balisage des Côtes de France," published in 1864; *Cornaglia*, "Sulla Verificazione degli Apparecchi Lenticolari per Fari," published in 1874; and *Christopher Nehls's* "Anhang" to the German translation of my book on Lighthouse Illumination, 1871. Nehls's translation was published at Hanover in 1878.

1. ABSTRACT of Professor SWAN'S Paper "ON NEW FORMS OF LIGHTHOUSE APPARATUS," in the Transactions of the Royal Society of Arts, 1867; furnished by the Author.

Totally reflecting lighthouse prisms, in order to distinguish them from other kinds of dioptric lighthouse apparatus, have hitherto been denominated *catadioptric.* For this word I have proposed to substitute *esoptric* (from ἔσοπτρον, a mirror), as sufficiently indicating the distinctive property of a *reflecting* or *mirror*-prism; and also to denote prisms as *monesoptric, diesoptric, tri-esoptric,* according as they reflect internally the intromitted rays of light, one, two, or three times.

General formulæ for the construction of Monesoptric Prisms are first investigated, for which, and for mathematical demonstrations generally as not falling within the scope of the present treatise, the reader is referred to the original memoir. These formulæ suffice for the construction of any monesoptric prism whatever, of which the following examples are here given:—

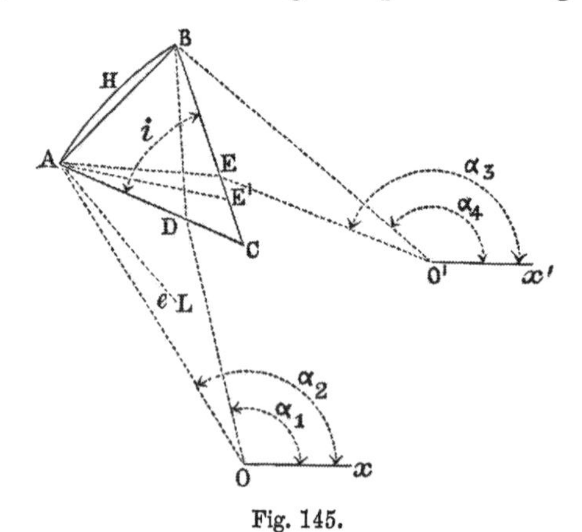

Fig. 145.

In the diagram, O is the luminous origin; A H B C a section of the prism perpendicular to its faces and in the same plane with O, and O A, O D, the extreme rays of the pencil of light incident on the face A C. The index of refraction being given it is required to determine the form of the section A H B C, where A C, B C are straight lines, and A H B a circular arc, so that the rays O D, O A, whose inclinations to the horizontal axis O x are given, shall be transmitted, after refraction at A C, reflection at A H B, and refraction at B C, in the directions B O′, E O′, whose inclinations to the horizontal are also given.

In the following formulæ the notation (that of the original memoir, slightly altered for the sake of simplicity) is as follows :—

$$\mathrm{DO}x = \alpha_1 \,,\ \mathrm{AO}x = \alpha_2 \,,\ \mathrm{EO'}x' = \alpha_3 \,,\ \mathrm{BO'}x' = \alpha_4 \,;$$

$$\mathrm{ACB} = i \,,\ \mathrm{BAC} = \mathrm{A} \,,\ \mathrm{ABC} = \mathrm{B} \,,\ \mathrm{LAO} = e \,;$$

and the angles of refraction at D and A, and of internal incidence at E and B, which are not shown in the diagram, r_1, r_2, r_3, r_4, respectively;

$\mathrm{BC} = a$, $\mathrm{AC} = b$, $\mathrm{AB} = c$, $\mathrm{AO} = d$, $\mathrm{AD} = f$, $\mathrm{BE} = h$, $\mathrm{BE'} = h'$, and the radius of curvature of $\mathrm{AHB} = \rho$.

Fresnel's Prisms.

(Chap. II. p. 65.)

In Fresnel's prisms the extreme ray OD is incident at C, and the intromitted rays AE, BD, are respectively coincident with the sides AC, BC. From these conditions it is shown that

$$\sin \tfrac{1}{2}\,(\alpha_1 - i) = \mu \cos i \,,$$

from which i is most readily obtained by trial: and the dimensions of the prism are then completely determined by the following formulæ; where it has been assumed that the prism, as in Fresnel's original construction, emits the rays BO′, EO′, *horizontally.*

$$\sin r_1 = \frac{1}{\mu} \sin \tfrac{1}{2}\,(i - \alpha_1) \,,$$

$$\sin r_2 = \frac{1}{\mu} \sin \tfrac{1}{2}\,(i + \alpha_1 - 2\,\alpha_2) \,;$$

$$\mathrm{A} = 90^\circ - \tfrac{1}{2}\,i + \tfrac{1}{4}\,(r_1 - r_2) \,,$$

$$\mathrm{B} = 90^\circ - \tfrac{1}{2}\,i - \tfrac{1}{4}\,(r_1 - r_2) \,,$$

$$\mathrm{C} = i \,;$$

$$b = d \sin (\alpha_2 - \alpha_1) \sec \tfrac{1}{2}\,(i - \alpha_1) \,,$$

$$a = b \sin \mathrm{A} \operatorname{cosec} \mathrm{B} \,,$$

$$c = b \sin \mathrm{C} \operatorname{cosec} \mathrm{B} \,,$$

$$\rho = \tfrac{1}{2}\,c \operatorname{cosec} \tfrac{1}{4}\,(r_1 - r_2) \,.$$

In this case, and in that also of the back prisms, the reader is referred to the original memoir for formulæ for the co-ordinates employed in their manufacture: and it has been thought advisable to add here numerical examples of both cases.

Example.

Given.	*Computed.*
$\alpha_1 = 34° \ 53'$,	$A = 33° \ 15' \ 0''$,
$\alpha_2 = 38° \ 47'$,	$B = 32° \ 9' \ 38''$,
$\alpha_3 = 180°$,	$C = 114° \ 35' \ 23''$,
$\alpha_4 = 180°$,	$a = 2{\cdot}7378$,
$d = 30$,	$b = 2{\cdot}6579$,
$\mu = 1{\cdot}54$ (*vide* note, p. 270).	$c = 4{\cdot}5405$,
	$\rho = 238{\cdot}7960$.

Back Prisms.

(Chap. II. p. 91.)

The form of these prisms is determined on the principle that among all the rays of the intromitted pencil, LA, which is incident *least* advantageously for total reflection, shall nevertheless be totally reflected at the "critical angle" θ, whose sine is equal to $\frac{1}{\mu_R}$; where μ_R is the index of refraction for the least refrangible rays of light. It is also assumed that the rays AO, EO′ are equally inclined to the faces AC BC of the prism.

The distance AO of the prism from the assumed focus in the flame, and the magnitude of the flame itself, being both supposed to be given, the angle LAO $= e$, can be found with sufficient accuracy. If, then,

$$D = e + \theta - \tfrac{1}{2} (\alpha_3 - \alpha_2),$$

and κ be a subsidiary angle assumed, so that

$$\tan \kappa = \frac{\mu_R + 1}{\mu_R - 1} \tan \tfrac{1}{2} D,$$

it is shown, that if ϕ denote the angle of incidence of the ray OA on AC, we shall have

$$\sin (\phi + e) = \frac{2 \mu_R}{\mu_R - 1} \cos \kappa \sin \tfrac{1}{2} D,$$

a relation from which, e being known, ϕ may be readily obtained; and lastly, it is found that

$$i = \pi + \alpha_2 - \alpha_3 - 2\phi .$$

In the case of a prism emitting focal rays horizontally, $\alpha_4 = \alpha_3 = 180°$; and the formulæ which completely determine its various dimensions are as follow:—

$$D = e + \theta + \alpha_2 - 90°,$$

$$\tan \kappa = \frac{\mu_R + 1}{\mu_R - 1} \tan \tfrac{1}{2} D ,$$

$$\sin (\phi + e) = \frac{2 \mu_R}{\mu_R - 1} \cos \kappa \sin \tfrac{1}{2} D ,$$

$$i = \alpha_2 - 2\phi ,$$

$$\sin r_1 = \frac{1}{\mu} \sin \tfrac{1}{2} (2 \alpha_1 - \alpha_2 - i) ,$$

$$\sin r_2 = \frac{1}{\mu} \sin \tfrac{1}{2} (\alpha_2 - i) ,$$

$$A = 90° - \tfrac{1}{2} i - \tfrac{1}{4} (r_2 - r_1) ,$$

$$B = 90° - \tfrac{1}{2} i + \tfrac{1}{4} (r_2 - r_1) ,$$

$$C = i .$$

$$f = d \sin (\alpha_2 - \alpha_1) \sec \tfrac{1}{2} (2\alpha_1 - \alpha_2 - i) ,$$

$$c = f \operatorname{cosec} \tfrac{1}{4} (2 i + 3 r_1 + r_2) \cos r_1 ,$$

$$b = c \sin B \operatorname{cosec} C ,$$

$$a = c \sin A \operatorname{cosec} C ,$$

$$h = c \sin \tfrac{1}{4} (2 i + r_1 + 3 r_2) \sec r_2 ,$$

$$\rho = \tfrac{1}{2} c \sin \tfrac{1}{4} (r_2 - r_1)$$

Light which would otherwise fall uselessly between D and C is intercepted by an adjacent prism or other optical agent, and superfluous glass may be cut off by a line joining D and E, but better by a line joining D with the point E′, at which the extreme ex-focal ray LA after reflection at A meets the side BC; so that none of the light may be lost, which otherwise the prism would usefully send out inclined downwards. If

$$\sin r'_2 = \frac{1}{\mu} \sin (+ e) ,$$

then BE′ or $h' = c \sin \tfrac{1}{4} (2 i + r_1 - r_2 + 4r'_2) \sec r'_2$.

Example.

Given.	*Computed.*
$a_1 = 114°$,	$A = 70° \; 8' \; 42''$,
$a_2 = 120°$,	$B = 71° \; 49' \; 10''$,
$a_3 = 180°$,	$C = 38° \; 2' \; 8''$,
$a_4 = 180°$,	$a = 10{\cdot}8659$,
$e = 6°$,	$b = 10{\cdot}9759$,
$d = 40$,	$c = 7{\cdot}1182$,
$\mu_R = 1{\cdot}50$,	$f = 5{\cdot}1031$,
$\mu = 1{\cdot}54$.*	$h' = 5{\cdot}8691$,
	$\rho = 243{\cdot}5763$.

Besides the new monesoptric prism here described, I have devised other forms in order to obtain deviations of light up to 180°,—two diesoptric and one triesoptric.

When Fresnel's or the new back prisms are employed for deviations exceeding 90°, a portion of the light will fall on a wrong side of the prism and be wasted, unless that side be screened or hidden from the luminous origin by an adjacent prism or other agent, as already noticed. In order to obtain additional facilities for designing lighthouse apparatus, I have proposed two forms of monesoptric prism, which admit of being freely exposed to the light without being in any way masked or *hidden*, a property which I have sought to express by the word *acryptic.*

It had long ago occurred to me that by an appropriate arrangement of the front prisms or lenticular zones, interspaces might be left through which to project back rays *at once in their proper directions*, and of such arrangements I have described and figured two designs. For details generally regarding these and various other matters, the reader is referred to the original memoir.

Dioptric Spherical Mirror.

(Chap. II. p. 87.)

Professor Swan's formulæ for this instrument have already been given.

* St. Gobain glass is now 1·53 ; Messrs. Chance's 1·51.

2. Fresnel's Lenticular Apparatus.

(Chap. II. p. 63.)

The following formulæ for the construction of Fresnel's Annular Lens and Fixed Light Apparatus, as furnished by Professor Swan, are, in accordance with the plan of this work, here presented without demonstration :—

The figure A B C D, whose side, C D, is an arc of a circle, and the other sides, of which A B is perpendicular to O x, are

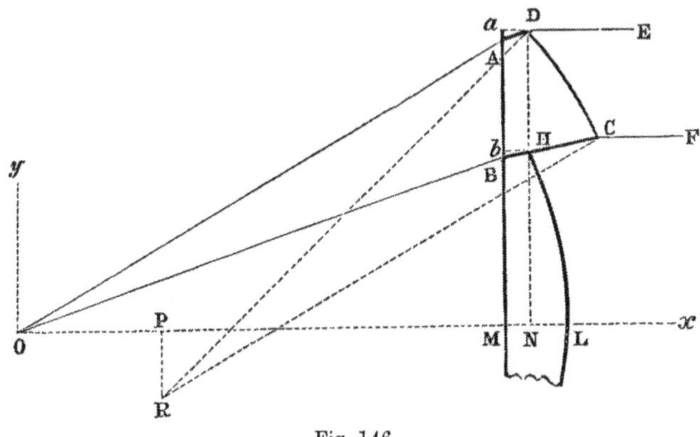

Fig. 146.

straight lines, represents the section of a refracting zone, which, according as it revolves about the horizontal axis, O x, or the vertical, O y, will generate a ring of an Annular Lens, or a refracting zone of a Fixed Light Apparatus.

Rays O A, O B, from the luminous origin at O, after refraction at A and B, proceed in the directions A D, B C; and after refraction at D and C, emerge in the directions D E, C F, parallel to O x.

For the rays O A, O B, let η, θ be the angles of incidence on the face A B, ϵ, ζ the corresponding angles of refraction, δ the angle of internal incidence at C, and α, β the angles of emergence at D and C.

Drawing D N parallel, and D A, H b perpendicular to A M, let

$$\text{M } a = a, \text{ M A} = c, \text{ M } b = b, \text{ M B} = d, \text{ O M} = f, \text{ M N} = t,$$

C R, the radius of curvature of the arc C D, $=\rho$, the co-ordinates of R the centre of curvature O P $= x$, P R $= y$, and μ the index of refraction, we have then—

$$\tan \eta = \frac{c}{f}, \qquad \tan \theta = \frac{d}{f},$$

$$\sin \epsilon = \frac{1}{\mu} \sin \eta, \qquad \sin \zeta = \frac{1}{\mu} \sin \theta,$$

$$\tan \alpha = \frac{\sin \eta}{\mu \cos \epsilon - 1}, \qquad \tan \beta = \frac{\sin \theta}{\mu \cos \zeta - 1},^* \qquad \sin \delta = \frac{1}{\mu} \sin \beta,$$

$$a = c + t \tan \epsilon, \qquad b = d + t \tan \zeta,$$

$$\rho = \frac{(a - b) \cos \zeta}{2 \sin \frac{1}{2} (\alpha - \beta) \cos \frac{1}{2} (\alpha - \beta + 2 \delta)},$$

$$x = f + t - \rho \cos \alpha, \qquad y = a - \rho \sin \alpha.$$

The radius of curvature, and co-ordinates of the centre of curvature, of the face C D of the prism A B C D being thus ascertained, its profile is completely determined.

If the ray O B coincide with O M, and O A with O B, then the generating profile A B C D will become B M L H, which, according as it revolves about O x or O y, will be that of the central disc of Fresnel's lens, or the middle zone of his Fixed Light Apparatus.

In this case the preceding formulæ at once reduce to the following :—

$$\tan \theta = \frac{d}{f}, \qquad \sin \zeta = \frac{1}{\mu} \sin \theta, \qquad \tan \beta = \frac{\sin \theta}{\mu \cos \zeta - 1},$$

* Subsidiary angles κ and λ being assumed, so that

$$\tan \kappa = \sin \eta, \text{ and } \tan \lambda = \sin \theta;$$

we may, if it be preferred, compute α and β by the formulæ—

$$\tan \alpha = \frac{\sin \kappa \sin \epsilon}{\sin (\kappa - \epsilon)}, \qquad \tan \beta = \frac{\sin \lambda \sin \zeta}{\sin (\lambda - \zeta)}.$$

$$b = d + t \tan \zeta, \qquad \rho = \frac{b}{\sin \beta},$$

$$x = f + t - \rho \cos \beta, \qquad y = 0.$$

In the construction here adopted, the lines A D, B H, will, in revolving, generate the surfaces of the joints by which the lenticular zone, whose section is A B C D, may be cemented to the adjoining portions of the apparatus; while in Fresnel's original design these joints were generated by the lines D *a*, H *b*; but the formulæ serve for either construction.

According to Fresnel's method, as given by Mr. Alan Stevenson in the notes appended to his "Account of the Skerryvore Lighthouse" (p. 250), the radius of curvature of the surface H L of the lens, or middle refracting zone, is obtained as follows:—

First, for rays near the point H, in the surface H L, there is found an expression for the radius of curvature equivalent to that already given. Next, for rays near the axis O L, the radius is obtained by the ordinary method for a small pencil incident directly on a lens. Denoting these two radii by ρ_1 and ρ_2, we shall then have

$$\rho_1 = \frac{b}{\sin \beta}, \qquad \rho_2 = (\mu - 1)(f + \frac{t}{\mu}).$$

A compromise between these values is then made by assuming, for the radius of the lens,

$$\rho = \tfrac{1}{2}(\rho_1 + \rho_2).$$

Example.—For a Fresnel's Fourth Order Apparatus, we may assume, in millimètres,

$$f = 150, \qquad c = 66, \qquad d = 40, \qquad \text{and } t = 3.$$

From these data, for glass whose index of refraction is $\mu = 1{\cdot}54$, the following dimensions have been computed by the preceding formulæ:—

For the surface C D,

$$\rho = 120{\cdot}14, \qquad x = 60{\cdot}46, \qquad y = -10{\cdot}62.$$

For the surface H L,

$$\rho_1 = 91{\cdot}01, \quad \rho_2 = 82{\cdot}05, \quad \rho = 86{\cdot}53, \quad x = 66{\cdot}47, \quad y = 0.$$

3. Differential Mirror.

(Chap. III. p. 104.)

Solution of Problem by Professor Tait.

The complete solution of your problem is easily given as an example of a beautiful method invented by Sir W. R. Hamilton.

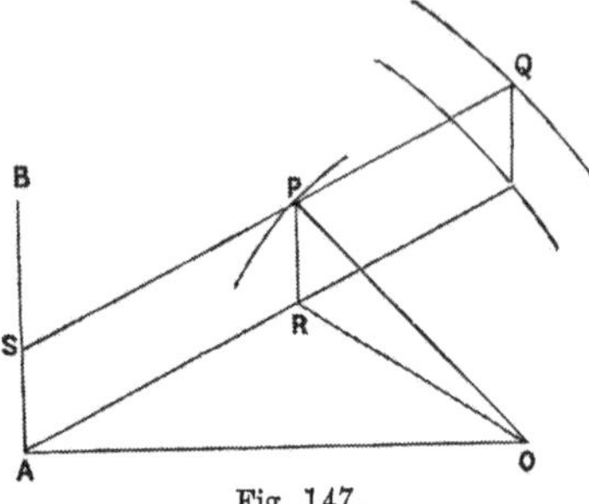

Fig. 147.

Suppose O to be the source of light, A B a vertical line from which the reflected rays are to diverge horizontally. Then after reflection they diverge in cylindrical surfaces of which A B is the axis. Suppose O P to be a ray reflected at P in the direction P Q. Then if Q P be produced backwards, it meets A B at right angles in (suppose), and our condition is O P + P Q = constant.
S P + P Q = constant. Hence the locus of P is such that
P O = constant, *i.e.*, the surface is such that the distance of each point in it from the axis A B differs from its distance from the luminous source O by a constant quantity.

The section by a vertical plane through O and A B is thus evidently a parabola with O as focus: that made by a horizontal plane through O a hyperbola of which A and O are the foci. One particular case of this hyperbola gives a horizontal straight line bisecting O A at right angles—so that a *parabolic cylinder* is one surface fulfilling the conditions.

The general equation above may be written—

$$\sqrt{(a-x)^2 + y^2} - \sqrt{x^2 + y^2 + z^2} = b$$

from which we get the parabolic section by putting $y = 0$, and the hyperbolic section by putting $z = 0$.

By giving z various values we get the sections of the surface made by various horizontal planes; and by giving various values to y those made by vertical planes parallel to the central parabolic section.

But it may more easily be constructed by the intersection of the spheres

$$x^2 + y^2 + z^2 = c^2$$

with the circular cylinders—

$$(a - x)^2 + y^2 = (b + c)^2,$$

and in various other ways, which might be given.

The values of a, b, and c, depend upon the diameter of the front lens, its distance from the light, and the *amount* of divergence required.

[Professor Tait has given another mode of dealing with this question in the Proc. R.S.E., 1870-71.]

REFERENCE TO PLATES.

CHAPTER I.

DESIGN AND CONSTRUCTION OF LIGHTHOUSE TOWERS.

CHAPTER II.

LIGHTHOUSE ILLUMINATION.

CHAPTER III.

AZIMUTHAL CONDENSING SYSTEM.

PLATES.

XIII. Intermittent condensing light for light periods of 20° and dark periods of 40° and 10°, p. 121.

XIV. Intermittent condensing light for light periods of 20° and dark periods of 80°, p. 122.

XV., XVI. Different forms of condensing spherical mirrors of unequal areas. In Plate XV. the light which would otherwise be lost on the mainland is sent seaward by a dioptric mirror. But to illuminate the sea up to the shore-line between A and F, the mirror is partially cut down to an extent in each of the arcs A C, C D, etc., inversely proportional to the lengths of the different landward ranges.

XXXII. The mirrors A B and C D are formed respectively by the revolution of parts of an ellipse A B and a hyperbola C D (having the same foci) round their common axis, one of the foci being in the centre of the flame. The rays falling on A B are reflected in the direction of the other focus, but being intercepted by C D, are returned through the flame and go to strengthen the arc opposite C D, while the arc subtended by A B is correspondingly weakened.

XXXIV. Intermittent condensing light, with spherical mirrors for light periods of 24°, and dark periods of 48°, p. 124.

XXXVI. Fixed condensing light of uniform intensity, changing from red to white without dark intervals, p. 124.

XVII. Condensing light for ships, shown in side elevation and middle horizontal section, in which A *a* A is 180° of fixed light apparatus, *pp*, straight condensing prisms, H, spherical mirror.

XVIII. Front elevation of above, in which *a* is 180° of fixed

PLATES.

light apparatus, B the straight condensing prisms, and O the gymbals in which the apparatus is hung, p. 117.

XX. Fixed condensing octant, as used at Buddonness, leading lights at River Tay, p. 109.

XXXVII. Condensing revolving light. Fig. 1 represents an existing apparatus which shows a fixed white light over the greater part of the horizon, with a red sector over a danger arc, while a larger sector is entirely lost towards the shore. In order to strengthen to the greatest extent the light over the danger arc (which has to be seen at a distance of 9 miles), there should be 3 sets of condensing prisms P, each of which spreads the light (which is at present lost) over the danger arc of 20°. Owing to want of room in the lightroom we cannot intercept and spread over the danger arc more than 83°, while 52° are still lost in the dark arc. This lost light may however be utilised by the following expedient :—If a portion of a dioptric spherical mirror E D be placed between the flame and the cylindric refractor of the main apparatus opposite the danger arc, the rays incident on it will be sent back through the flame, so as to strengthen that arc, partly directly and partly by the condensing prisms. But this arrangement will manifestly greatly weaken the power of the main apparatus in the seaward direction E D. To restore uniformity of strength over this weakened arc, the unutilised part of the light going landwards must be deflected by twin (Chap. III.) and condensing prisms P′, so as to spread it over the other seaward arc E D. The strength of the light, less absorption, etc., over the danger arc will thus be increased about five times by the condensing prisms P, and twice by the spherical mirror, while that over the seaward are will still be maintained of uniform strength by the prisms P′. Finally, by the substitution of moving screens A B, B C (Chap. IV.), for the red shades, the danger arc will be made to show an intermittent white light, and thus be distinguished in character from the rest of the apparatus. By using white instead of red, as at present, the power will be increased about 4 times, so that by this means, and the condensing prisms P, and the spherical mirror E D, the danger arc will be strengthened (less absorption, etc.) about 28 times.

Fig. 2 shows a mode of utilising the light which is not intercepted by the spherical mirror, and which would be lost by passing landwards at stations where only a certain seaward sector has to be illuminated. In the drawing, which represents a revolving light, the parts marked R P L B revolve, while those marked F E M are stationary. E is a portion of a holophote which projects upwards, in an angle inclined to the vertical, the rays which it intercepts. These rays fall on fixed expanding prisms F, at the top of the apparatus, which cause them to spread through half of the circle, in azimuth.

CHAPTER IV.

DISTINCTIVE CHARACTERISTICS OF LIGHTS.

PLATES.

XXII. Dr. Hopkinson's group flashing light, p. 136.

XXII. Figs. 1 and 2. Drawing in illustration of experiments on coloured lights, p. 140.

XXIV. Fig. 1. Reversing light, p. 129.

XXXV. Mr. Brebner's form of group-flashing apparatus, p. 136.

CHAPTER V.

ILLUMINATION OF BEACONS AND BUOYS.

PLATES.

XXIV. Pintsch's floating buoy, lighted with gas, p. 169.

XXVIII. Different forms of buoys suitable for different localities, p. 177.

XXIX. Courtenay's automatic signal buoy, p. 173.

XXX. Malleable-iron beacon, generally employed in Scotland, p. 177.

XXXI. Cast-iron beacon employed in Scotland at rocks exposed to heavy seas, p. 177.

XXXII. Dipping lights for covering shoals and sunk rocks, p. 153.

CHAPTER VI.

ELECTRIC LIGHT.

XIX. Mr. Brebner's *Double Lens.*—In the upper diagrams f_1 is the focus for parallel horizontal rays for the lower part, and f_2 the same for the upper part of the lens; the flame has been moved forward from f_1 to the position shown, in order to give the required horizontal divergence; at the same time, by cutting away and lowering the upper part of the lens to an extent $f_1\ f_2$, it is ensured that the rays of greatest vertical divergence from the upper half of the unaltered lens are sent to the horizon by the same as modified. The light from the lower half is strongest in the direction of the horizon, and becomes gradually weaker to an observer approaching the lighthouse; the converse holds for the light from the upper part of the lens, so that in the plane of the middle vertical section of the modified lens, and approximately in the

PLATES.

vertical planes at either side of this, an equally intense light is directed to all points on the sea between the horizon and the inner boundary of the illuminated surface.

In connection with this lens Mr. Brebner designed to give such a form to the upper and lower totally reflecting prisms as would strengthen the light sent to the horizon or any other part of the sea where desired.

Differential Lens, p. 187.—The lower diagrams show in elevation and horizontal section a differential lens for the electric light. In vertical section the inner surface of the lens is straight; in horizontal section the centre of curvature of the inner surface is taken so as to give to the light the required horizontal divergence.

XXXII. Fig. 4. Lens for electric light, as designed for the Trinity House by Mr. Chance.

CHAPTER VII.

LIGHTHOUSE STATISTICS.

XXXIII. Graphic Table, deduced from M. Allard's work, for the ranges of lights, p. 202.

CHAPTER IX.

MISCELLANEOUS SUBJECTS CONNECTED WITH LIGHTHOUSES.

XI., XII. Mr. Alan Stevenson's diagonal lantern of bronze, and Mr. Douglass's helical lantern of steel, pp. 75 and 223.

XXIV. Different forms of photometers, p. 244.

CHAPTER X.

FOG-SIGNALS.

INDEX.

THE END.

Printed by R. & R. Clark, *Edinburgh.*

PLATE I

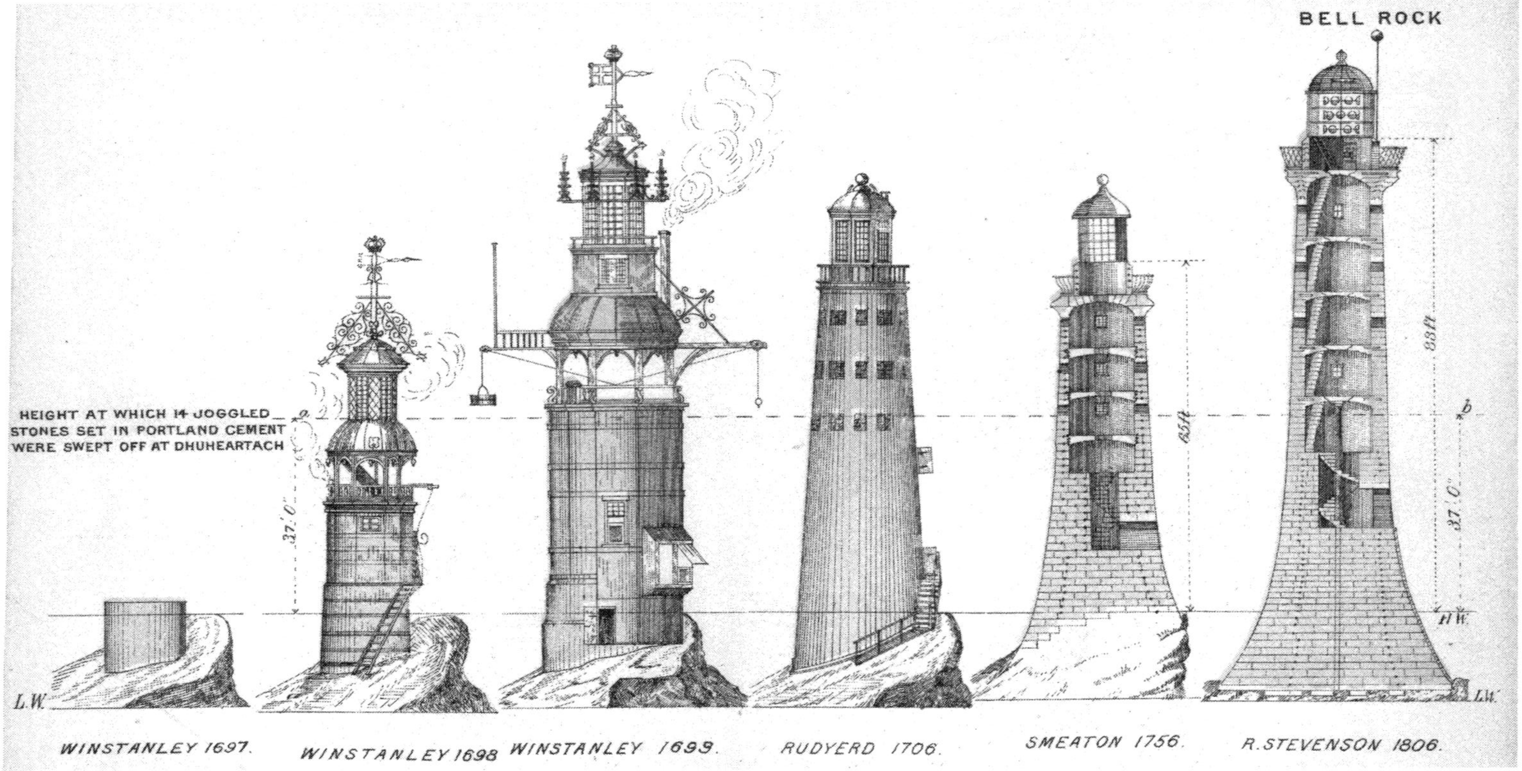

BELL ROCK

HEIGHT AT WHICH 14 JOGGLED STONES SET IN PORTLAND CEMENT WERE SWEPT OFF AT DHUHEARTACH

a

37' 0"

b

85 ft

88 ft

37' 0"

H.W.

L.W.

L.W.

WINSTANLEY 1697. WINSTANLEY 1698 WINSTANLEY 1699. RUDYERD 1706. SMEATON 1756. R. STEVENSON 1806.

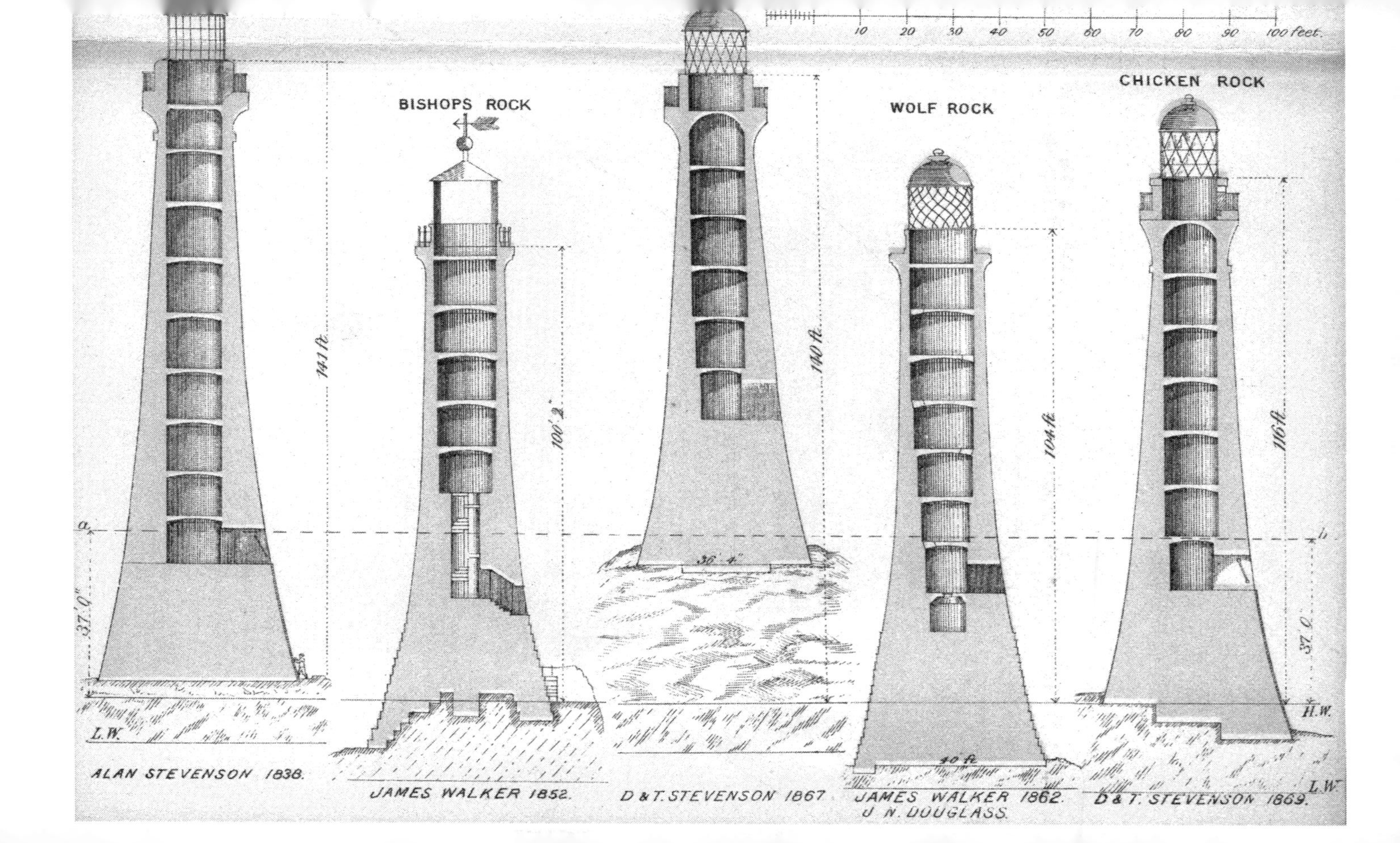
10 20 30 40 50 60 70 80 90 100 feet.
BISHOPS ROCK
WOLF ROCK
CHICKEN ROCK
141 ft.
37' 0"
a
L.W.
H.W.
100' 2"
140 ft.
36' 4"
104 ft.
40 ft.
116 ft.
37' 0
b
ALAN STEVENSON 1838.
JAMES WALKER 1852.
D & T. STEVENSON 1867.
JAMES WALKER 1862.
J. N. DOUGLASS.
D & T. STEVENSON 1869.

PLATE II

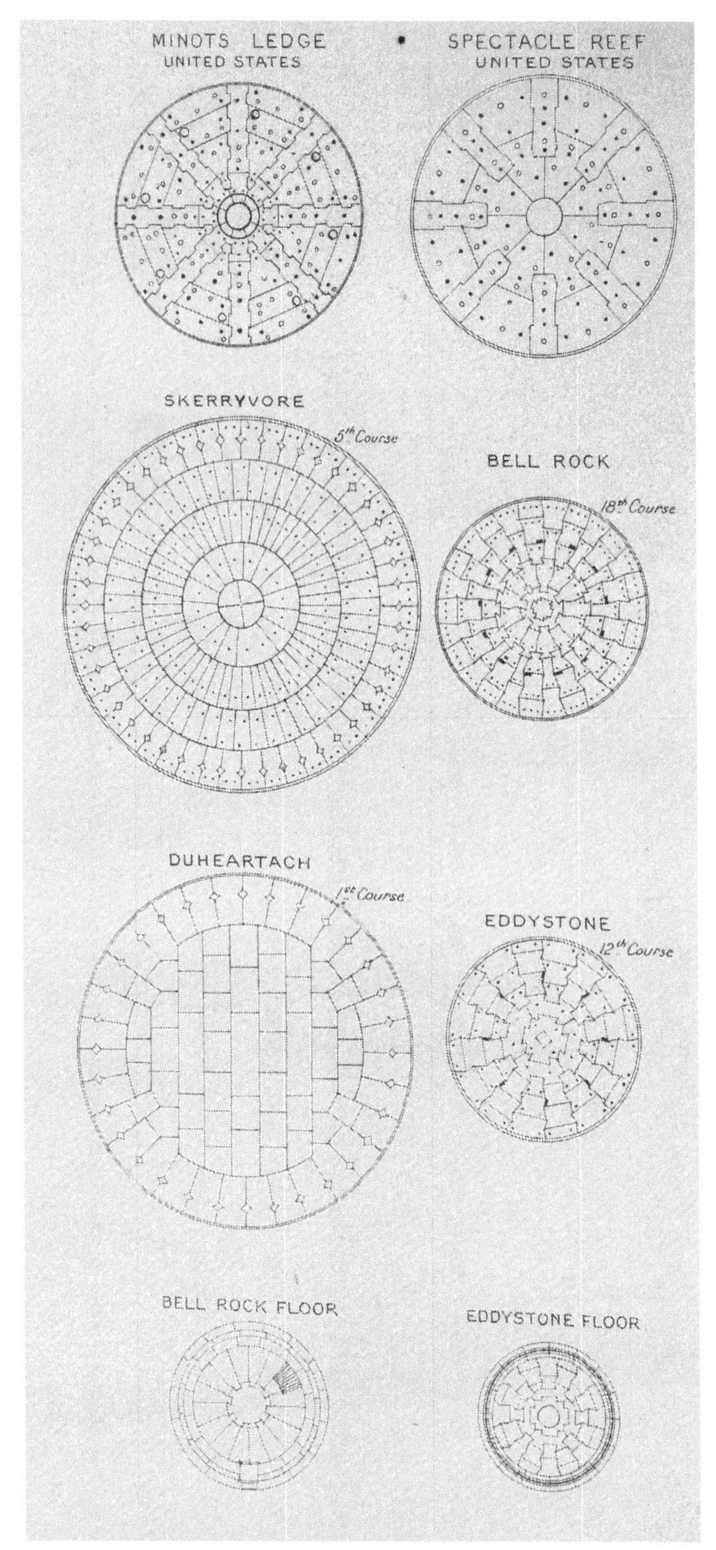
MINOTS LEDGE
UNITED STATES
SPECTACLE REEF
UNITED STATES
SKERRYVORE
5th Course
BELL ROCK
18th Course
DUHEARTACH
1st Course
EDDYSTONE
12th Course
BELL ROCK FLOOR
EDDYSTONE FLOOR

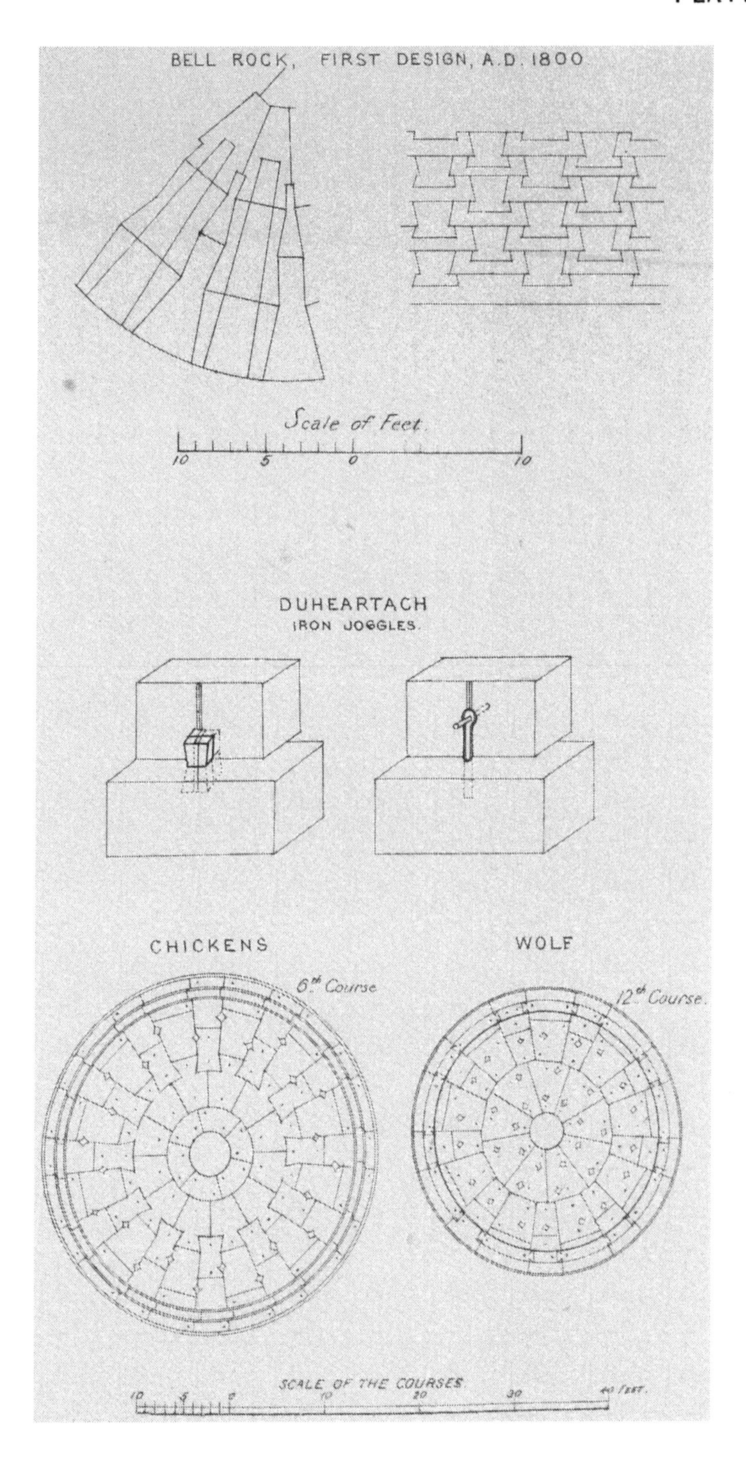
BELL ROCK, FIRST DESIGN, A.D. 1800
Scale of Feet.
10 5 0 10
DUHEARTACH
IRON JOGGLES.
CHICKENS
WOLF
6th Course
12th Course.
SCALE OF THE COURSES.
10 5 0 10 20 30 40 FEET.

TEMPORARY BARRACK
Used at the erection of the
BELL ROCK & SKERRYVORE
LIGHTHOUSES.

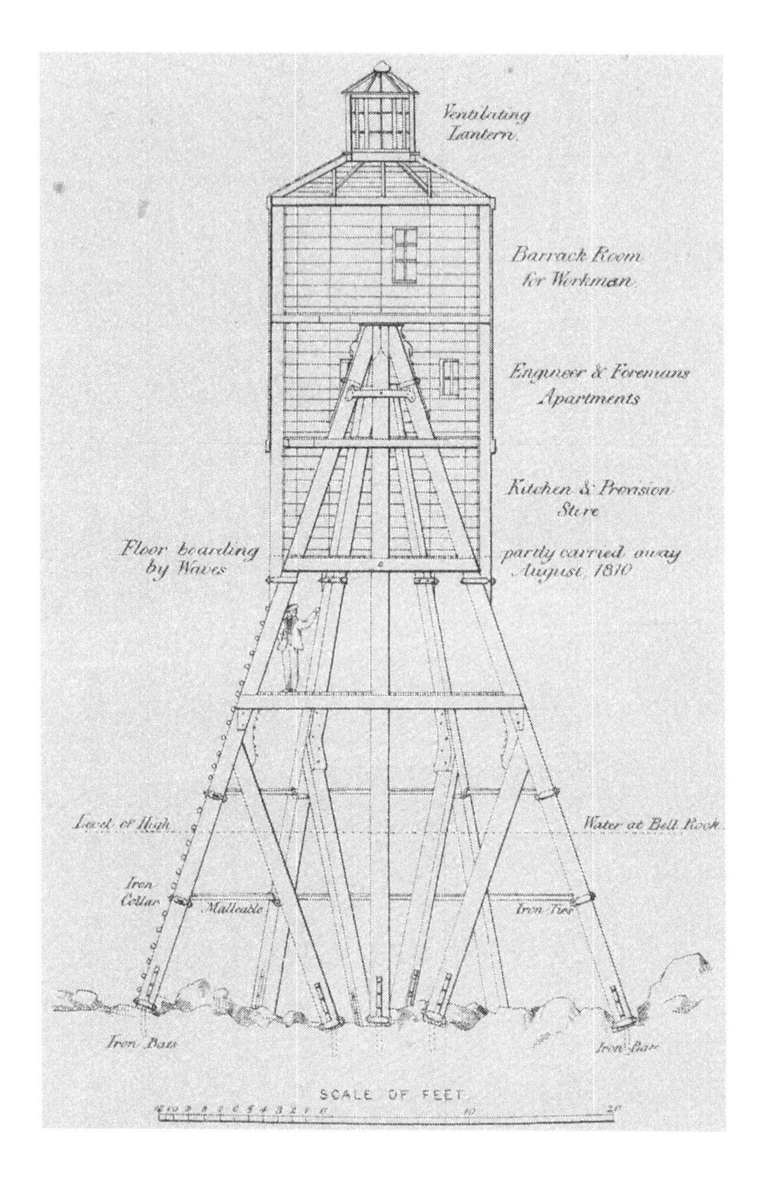

DHUHEARTACH

PLATE V

WORKMENS BARRACK ON THE ROCK.

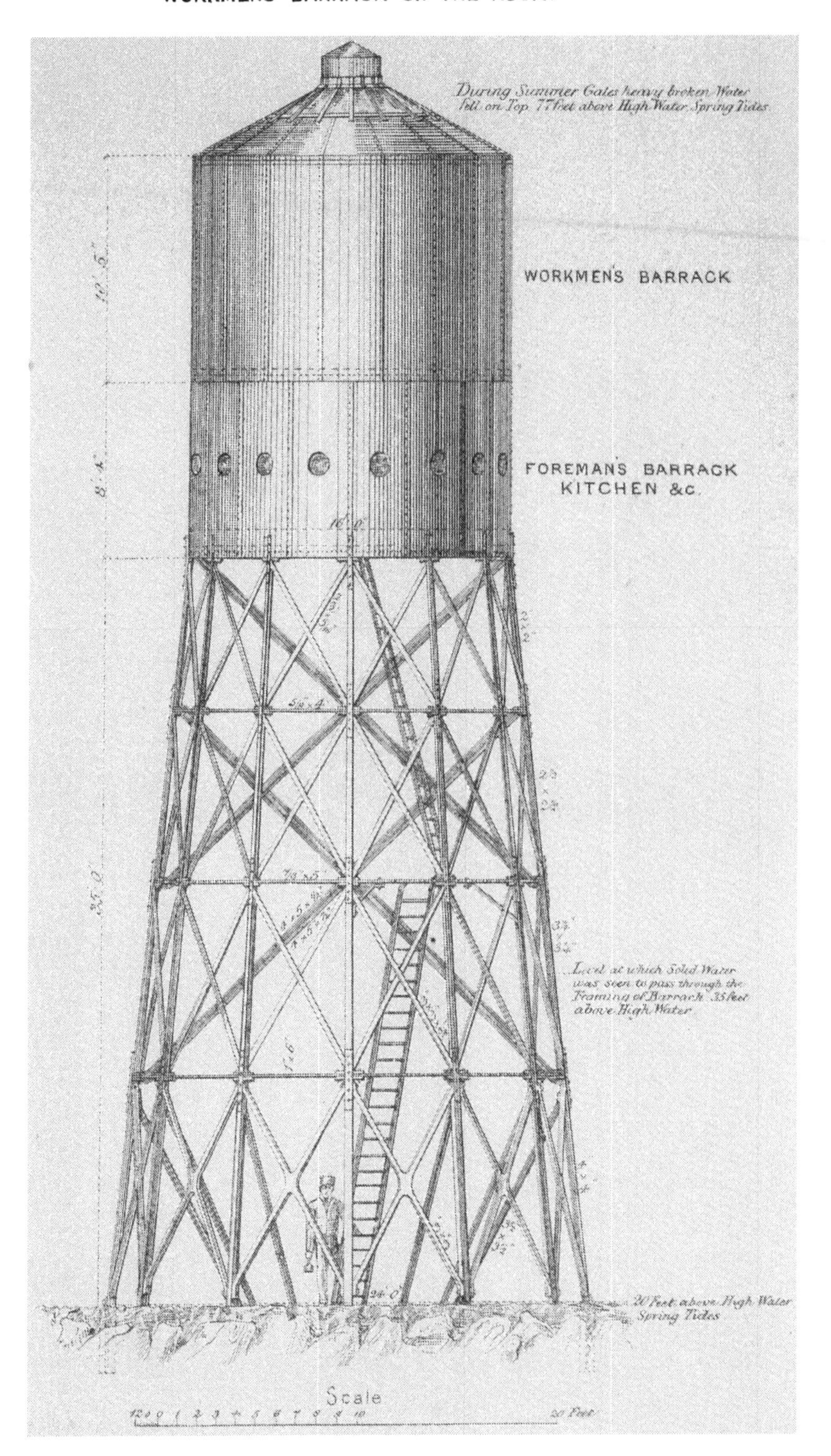

PLATE VI

FASTNET LIGHTHOUSE

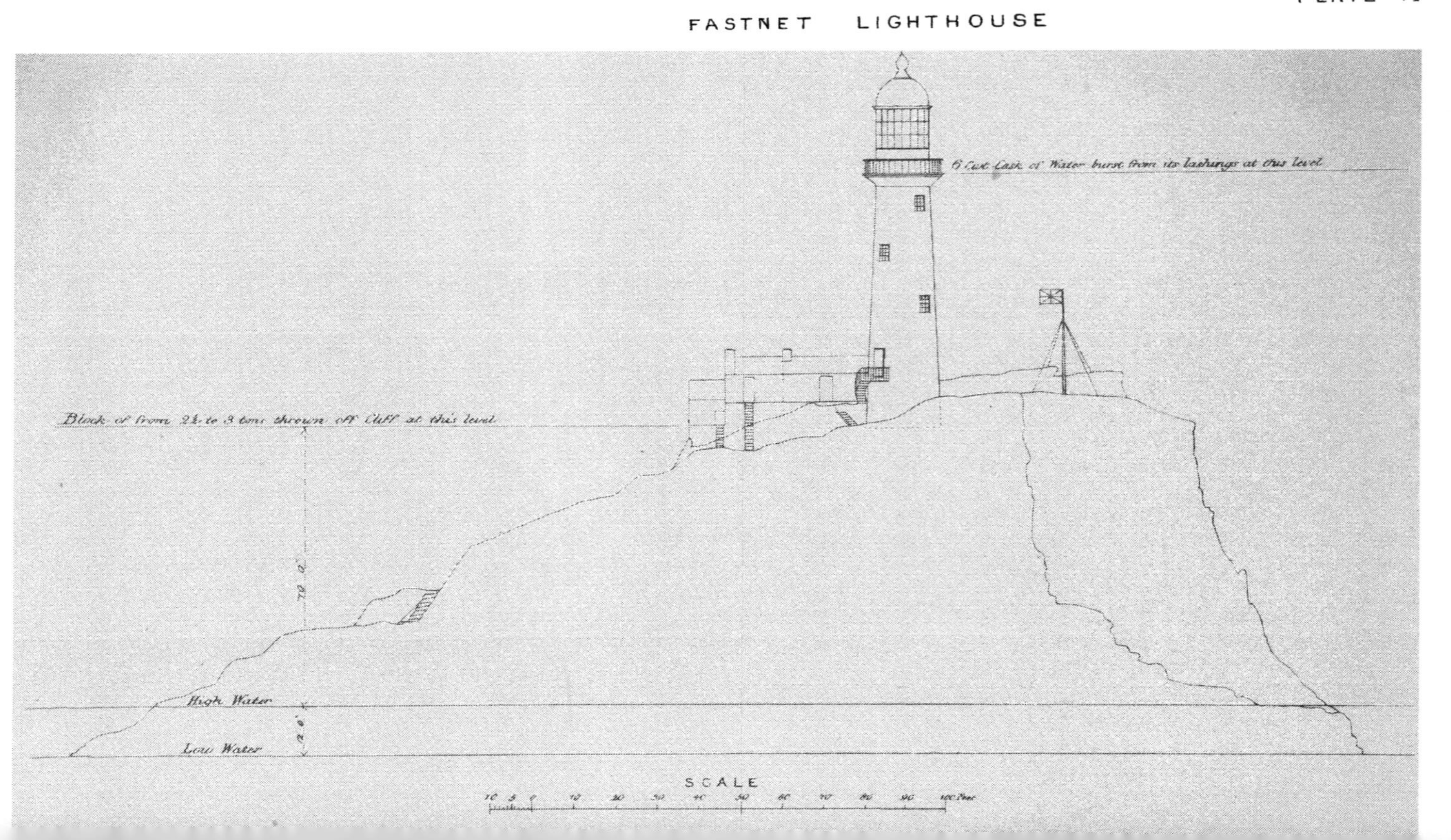

CRAIGHILL'S CHANNEL RANGE OR LEADING LIGHTS
CHESAPEAKE BAY — THE LOW LIGHT

LIGHT-HOUSE AT SPECTACLE REEF.
LAKE HURON
(In an Ice-Floe.)

PLATE IX

ARDNAMURCHAN LIGHTHOUSE.

ARDNAMURCHAN LIGHTHOUSE.

PLAN

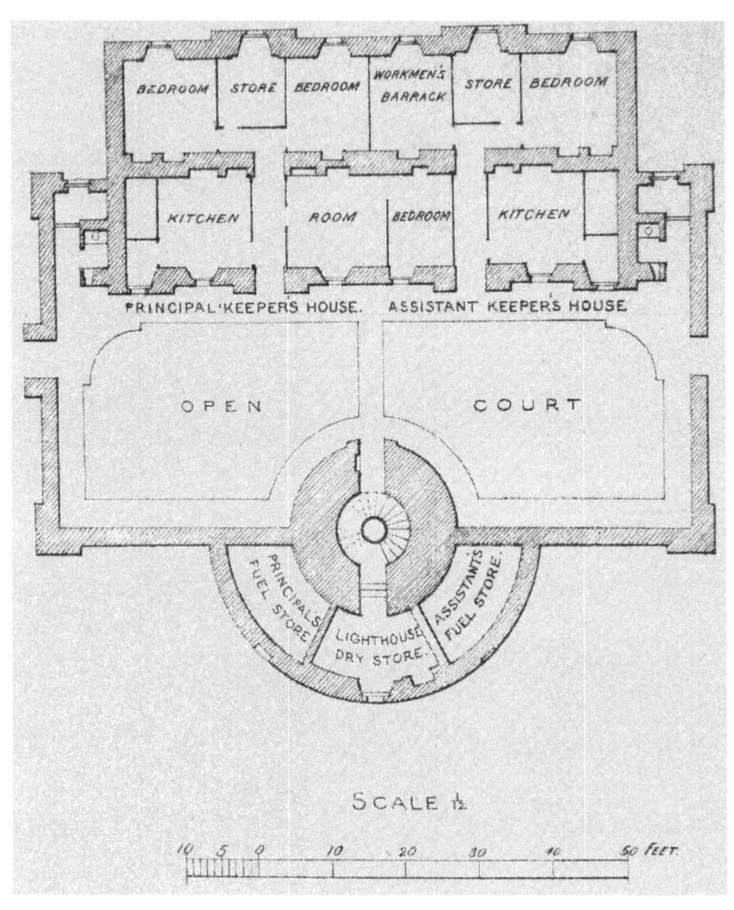

PLATE XI.

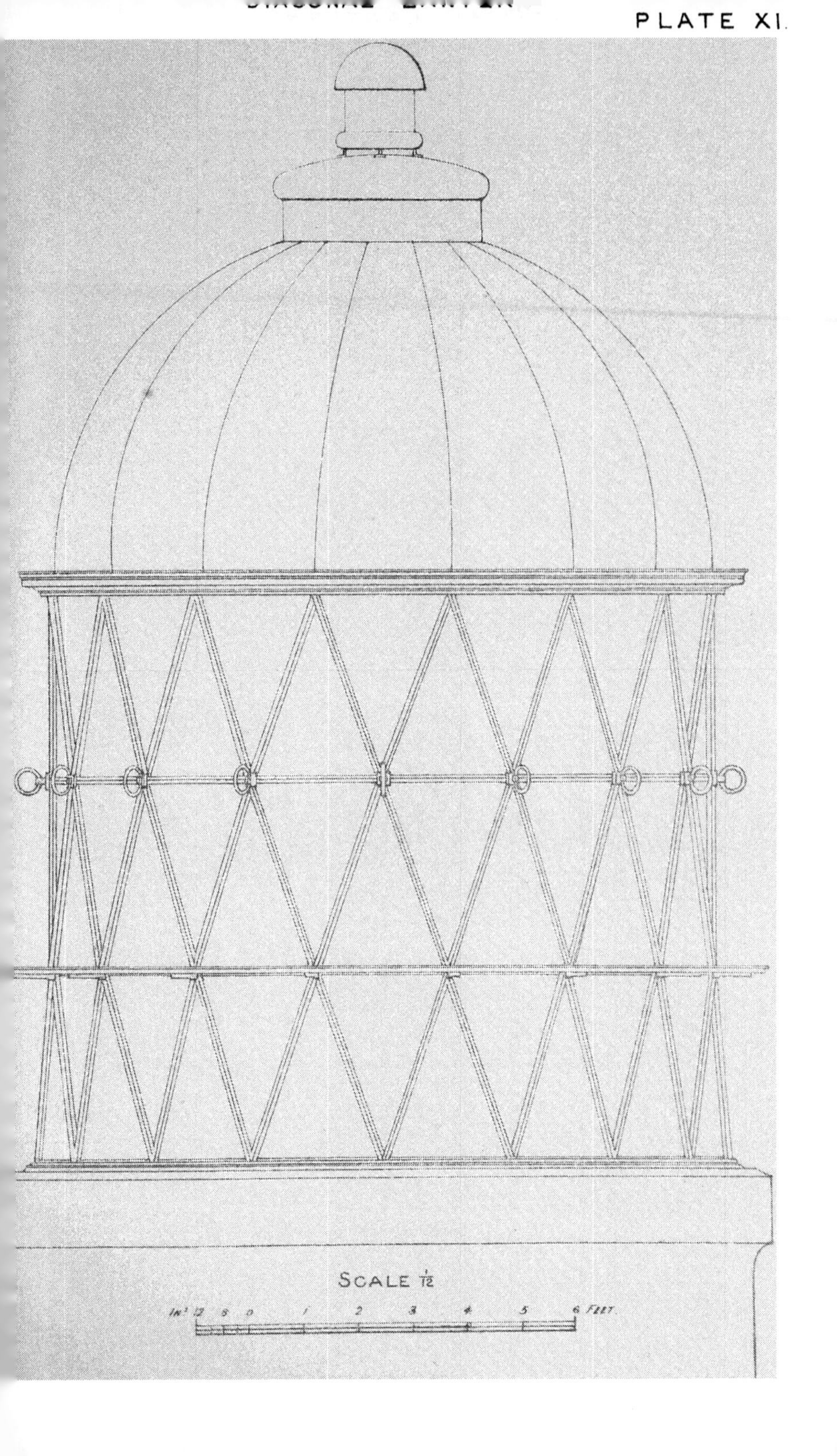

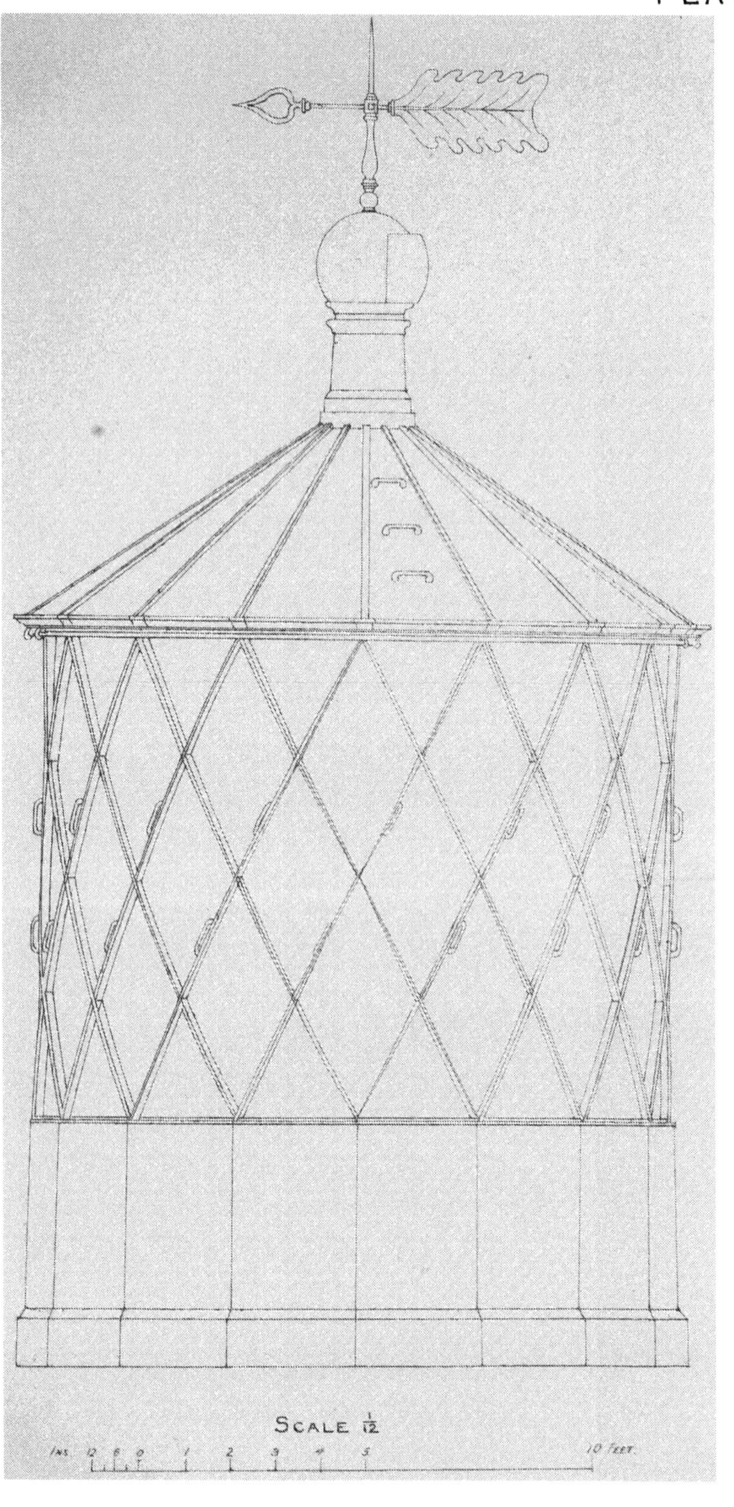
SCALE $\frac{1}{12}$
INS. 12 6 0 1 2 3 4 5 10 FEET.

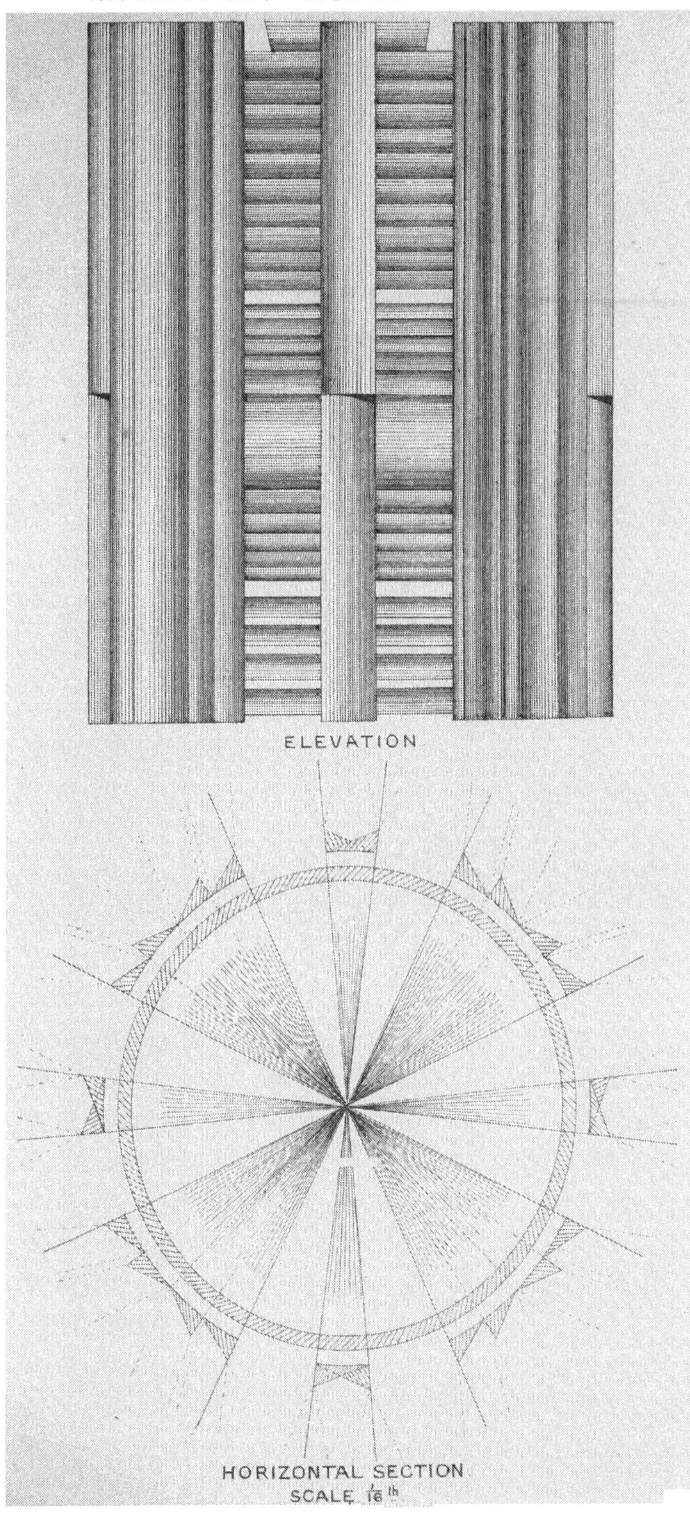
ELEVATION
HORIZONTAL SECTION
SCALE $\frac{1}{16}$th

AZIMUTHAL CONDENSING SYSTEM.

INTERMITTENT LIGHTS

PLATE X

ELEVATION.

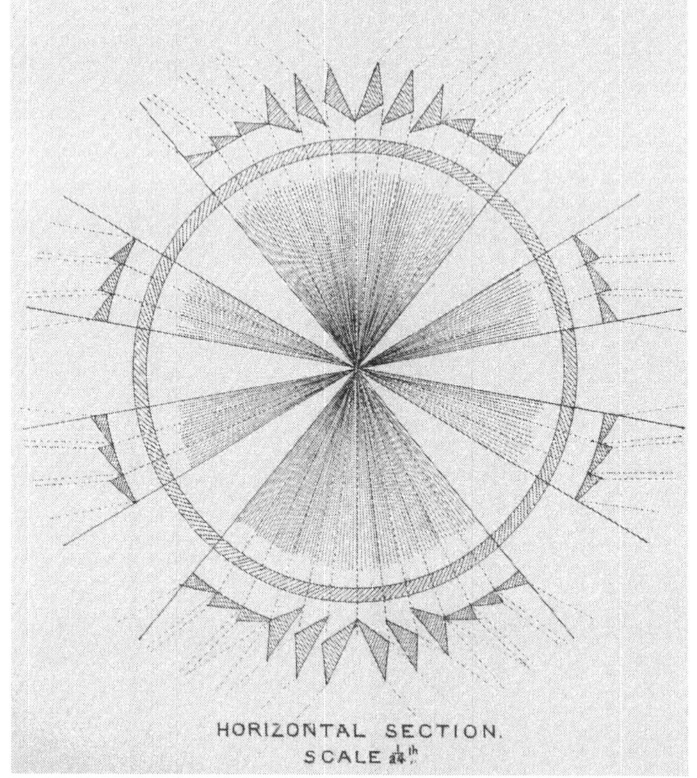

HORIZONTAL SECTION.

SCALE $\frac{1}{24}$th.

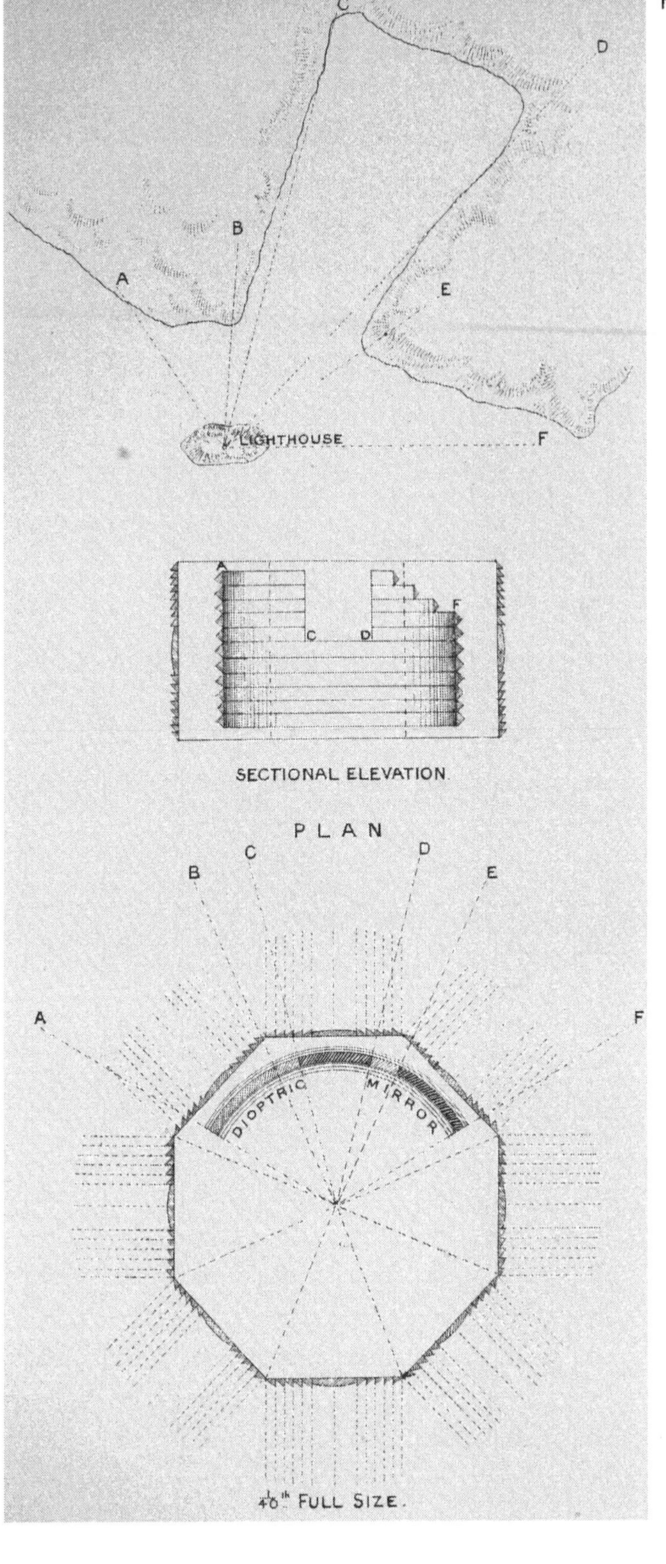
C
D
B
A
E
LIGHTHOUSE
F
A
C
D
F
SECTIONAL ELEVATION.
PLAN
B
C
D
E
A
F
DIOPTRIC
MIRROR
$\frac{1}{40}$th FULL SIZE.

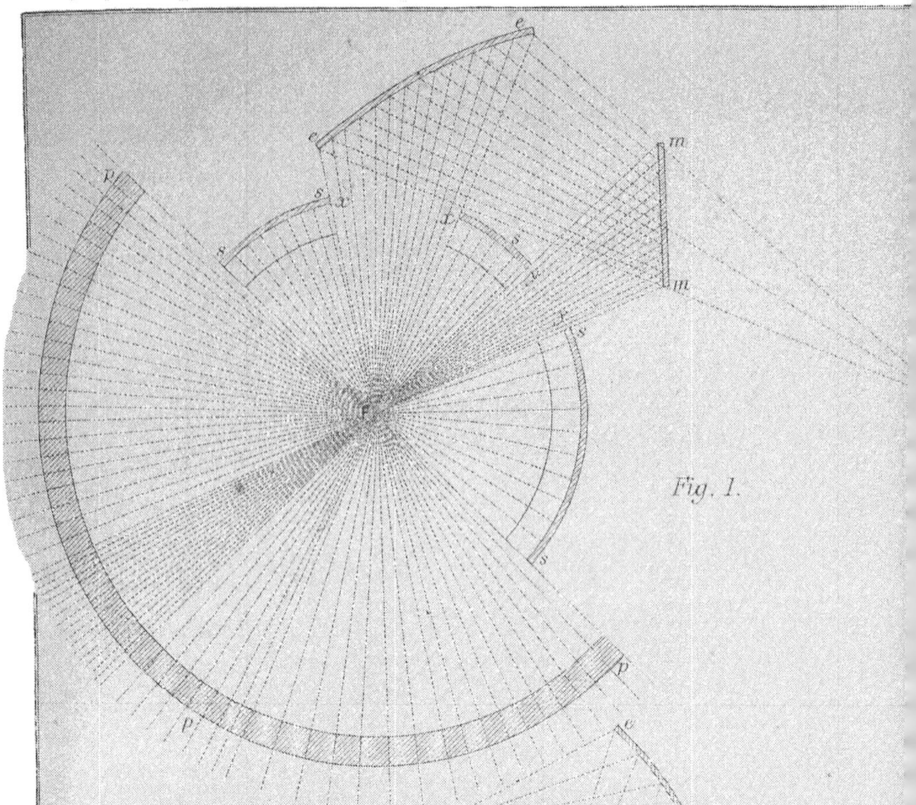

Fig. 1.

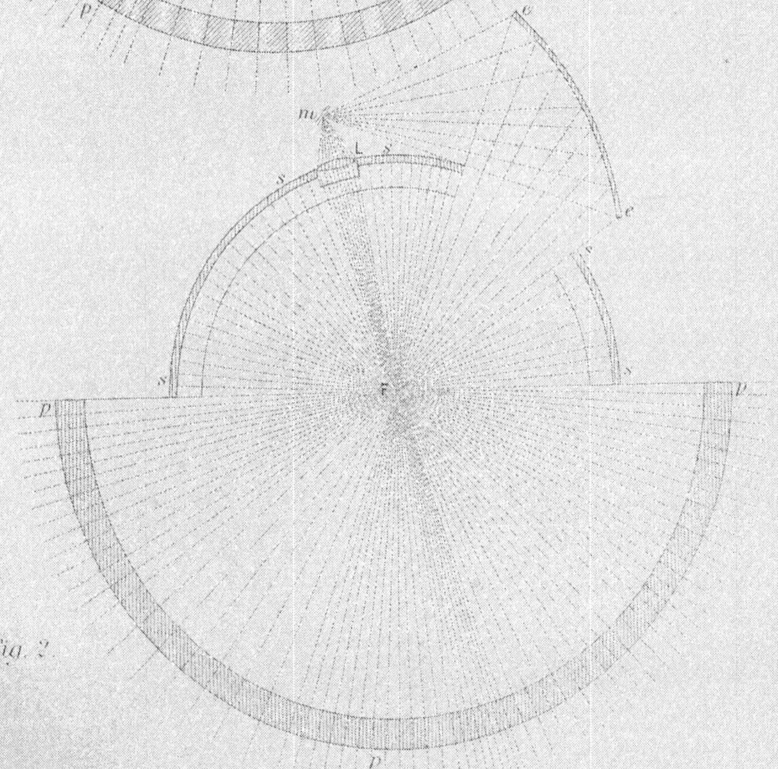

Fig. 2.

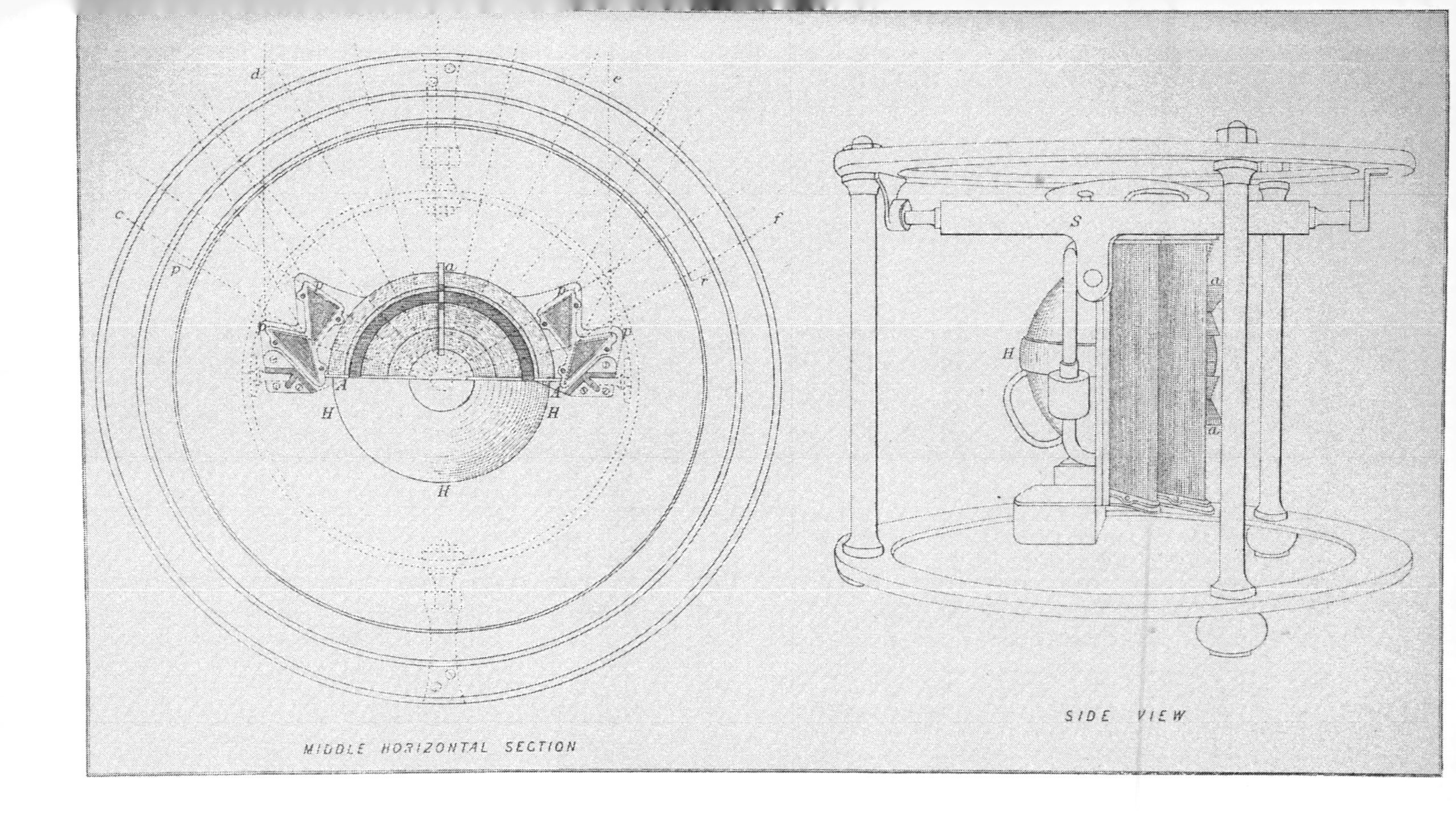
MIDDLE HORIZONTAL SECTION
SIDE VIEW

FRONT ELEVATION

DOUBLE LENS FOR ELECTRIC LIGHT.

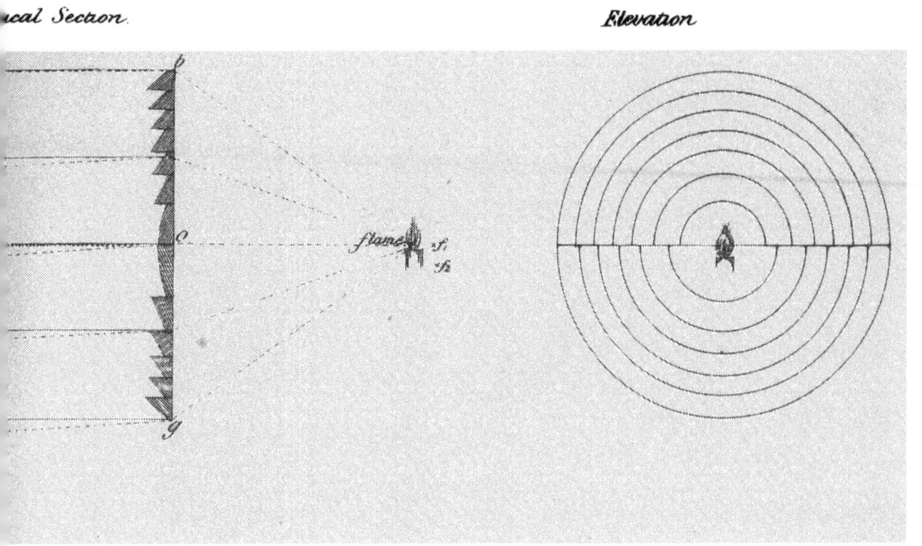

DIFFERENTIAL LENS FOR ELECTRIC LIGHT.

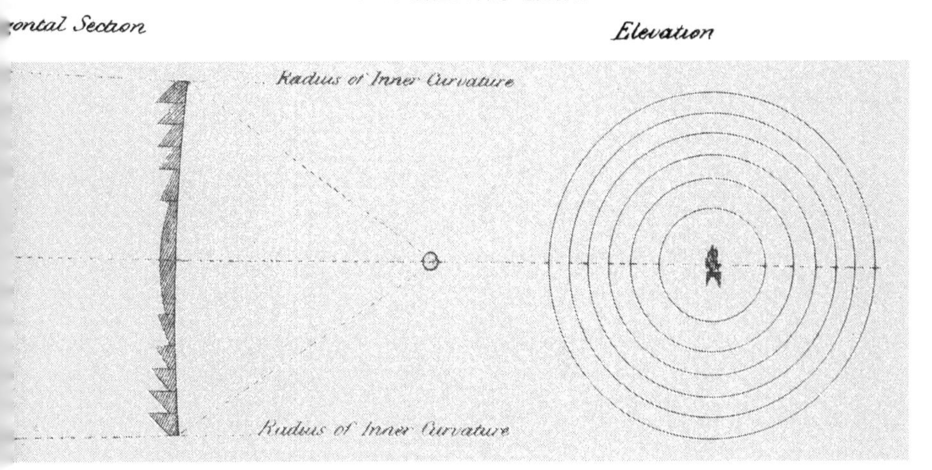

SCALE $\frac{1}{12}$TH.

www.ingramcontent.com/pod-product-compliance
Ingram Content Group UK Ltd.
Pitfield, Milton Keynes, MK11 3LW, UK
UKHW040658180125
453697UK00010B/250